SpringerBriefs in Applied Sciences and Technology

Series editor

Janusz Kacprzyk, Polish Academy of Sciences, Systems Research Institute, Warsaw, Poland

SpringerBriefs present concise summaries of cutting-edge research and practical applications across a wide spectrum of fields. Featuring compact volumes of 50–125 pages, the series covers a range of content from professional to academic.

Typical publications can be:

- A timely report of state-of-the art methods
- An introduction to or a manual for the application of mathematical or computer techniques
- A bridge between new research results, as published in journal articles
- A snapshot of a hot or emerging topic
- An in-depth case study
- A presentation of core concepts that students must understand in order to make independent contributions

SpringerBriefs are characterized by fast, global electronic dissemination, standard publishing contracts, standardized manuscript preparation and formatting guidelines, and expedited production schedules.

On the one hand, **SpringerBriefs in Applied Sciences and Technology** are devoted to the publication of fundamentals and applications within the different classical engineering disciplines as well as in interdisciplinary fields that recently emerged between these areas. On the other hand, as the boundary separating fundamental research and applied technology is more and more dissolving, this series is particularly open to trans-disciplinary topics between fundamental science and engineering.

Indexed by EI-Compendex and Springerlink.

More information about this series at http://www.springer.com/series/8884

Neftali L.V. Carreño · Ananda M. Barbosa
Bruno S. Noremberg · Mabel M.S. Salas
Susana C.M. Fernandes · Jalel Labidi

Advances in Nanostructured Cellulose-based Biomaterials

 Springer

Neftali L.V. Carreño
Graduate Program in Materials Science
and Engineering, Technology
Development Center
Federal University of Pelotas
Pelotas, RS
Brazil

Ananda M. Barbosa
Graduate Program in Materials Science
and Engineering, Technology
Development Center
Federal University of Pelotas
Pelotas, RS
Brazil

Bruno S. Noremberg
Graduate Program in Materials Science
and Engineering, Technology
Development Center
Federal University of Pelotas
Pelotas, RS
Brazil

Mabel M.S. Salas
Graduate Program in Dentistry
Federal University of Juiz de Fora
Governador Valadares, MG
Brazil

Susana C.M. Fernandes
Institut des Sciences Analytiques et de
Physico-Chimie pourl'Environnement et
les Matreriaux, UMR 5254
CNRS/Univ Pau & Pays Adour
Pau
France

Jalel Labidi
Chemical and Environmental Engineering
Department
University of the Basque Country
San Sebastian
Spain

ISSN 2191-530X ISSN 2191-5318 (electronic)
SpringerBriefs in Applied Sciences and Technology
ISBN 978-3-319-58156-9 ISBN 978-3-319-58158-3 (eBook)
DOI 10.1007/978-3-319-58158-3

Library of Congress Control Number: 2017940610

Printed on acid-free paper

This Springer imprint is published by Springer Nature
The registered company is Springer International Publishing AG
The registered company address is: Gewerbestrasse 11, 6330 Cham, Switzerland

Contents

Contents

Abstract

Nanostructured cellulose, a unique natural material isolated from plants and some animals or biosynthesized by bacteria, has served society in several fields of life, including medicine and biotechnology. Thanks to its nontoxic, biocompatible and biodegradable nature, nanostructured cellulose gives a unique opportunity to be used in the design and preparation of biomaterials.

This book was conceived to provide an overview of the advances and relevant research in nanostructured cellulose in the development of biomaterials for delivery systems and tissue engineering. An important advantage of nanocellulose is its nanometric size, which provides high surface area and high capacity to interact with other materials. This manuscript also describes how the nanometric properties of cellulose can be improved to increase their range of application. In Future Perspectives, new technologies in the biomedical field are portrayed.

Chapter 1
Advances in Nanostructured Cellulose-based Biomaterials

1.1 Introduction

Cellulose is one of the most important natural polymers with high availability throughout the planet. It is present in trees, plants, fruits, barks and leaves representing the main structural element of the cell wall of plant tissues.

Structurally, cellulose is a carbohydrate polymer composed of repeating β-D-glucopyranose molecules that are covalently linked through acetal functions between the equatorial OH group of C4 and the C1 carbon atom (β-1,4-glucan). Cellulose is a linear homopolysaccharide (Fig. 1.1) with a large number of hydroxyl groups (three per anhydroglucose (AGU) unit) present in the thermodynamically preferred $^4C1$ conformation (Klemm et al. 2005).

During the biosynthesis of cellulose, the cellulose molecules pile onto each other forming microfibrils (diameter of 2–30 nm), which form both crystalline regions (densely and firmly packed and their hydroxyl groups become unavailable) and amorphous regions (more available with high reactivity towards chemical species). The microfibrils aggregate into fibrils with diameters of 30–100 nm and lengths of 100–500 μm. Finally, these fibrils aggregate into cellulose fibres with diameters of 100–400 nm and lengths of 0.5–4 mm (Klemm et al. 1998).

Thousands of years prior to the discovery of the "sugar of the plant cell wall", cellulose was used in different forms namely wood, cotton and plant fibres as energy source, building materials, and fibres for clothing (Klemm 2005).

More recently, using effective chemical, mechanical methods, is possible to obtain nanosize cellulose namely nanocrystals and nanofibrils from the conventional cellulose fibers. These nanostructured cellulose materials have been used as reinforcing phase in a polymeric matrix or as a matrix to make sustainable, mechanical high performance, functional and bioactive sophisticated biomaterials for nanomedicine (Siro 2010; Klemm et al. 2011; Kolakovic 2012). Nanostructured cellulose is widely applied in medical implants, tissue engineering, drug delivery,

© The Author(s) 2017
N.L.V. Carreño et al., *Advances in Nanostructured Cellulose-based Biomaterials*,
SpringerBriefs in Applied Sciences and Technology,
DOI 10.1007/978-3-319-58158-3_1

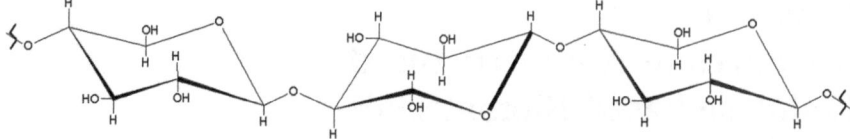

Fig. 1.1 Chemical structure of cellulose

wound healing, cardiovascular applications, and other medical application (Jorfi and Foster 2014).

The main goal of this book is to review the recent scientific advances and future perspectives regarding these sophisticated nanostructured cellulose-based materials targeted for biomedical applications.

1.2 Nanocellulose Existing Forms

Nanostructured cellulose has been grouped in three types, and new terms have been used for the same material: (1) cellulose nanocrystals (CNCs), also known as nanocrystalline cellulose (NCC) and cellulose nanowhiskers (CNWs); (2) cellulose nanofibrils (CNFs), also called nano-fibrillated cellulose (NFC); and (3) bacterial cellulose (BC), also identified as microbial cellulose or biocellulose (Klemm et al. 2011; Abitbol et al. 2016). Some examples of approaches that have been developed to achieve these different nanocellulose materials are described in the next section.

1.2.1 Isolation Methods to Obtain Nanostructured Cellulose

As already mentioned, cellulose is mainly obtained from plants. Thus, cellulose can be extracted from lignocellulosic sources like softwood, hardwood and from several non-woody plants namely flax, hemp, jute, sisal, abaca, cotton, among others (Alila et al. 2013).

Pulping is an example of procedures to converts lignocellulosic material into a fibrous material. The processes can be mechanical, chemical, or a combination of the two methods (Fig. 1.2). The main mechanical processes to extracted pulps are thermo mechanical pulp (TMP) and groundwood pulp (GW). Basically, mechanical forces are used to produce the pulp, with or without the use of high temperatures. The three major chemical pulping processes are: kraft, soda, and sulphite. In this case, chemical products and heat are applied to dissolve lignin and isolate the fibres.

As represented in Fig. 1.2, from wood pulp cellulose fibres is possible to isolate microfibrillated cellulose (MFC) by mechanical disintegration and cellulose

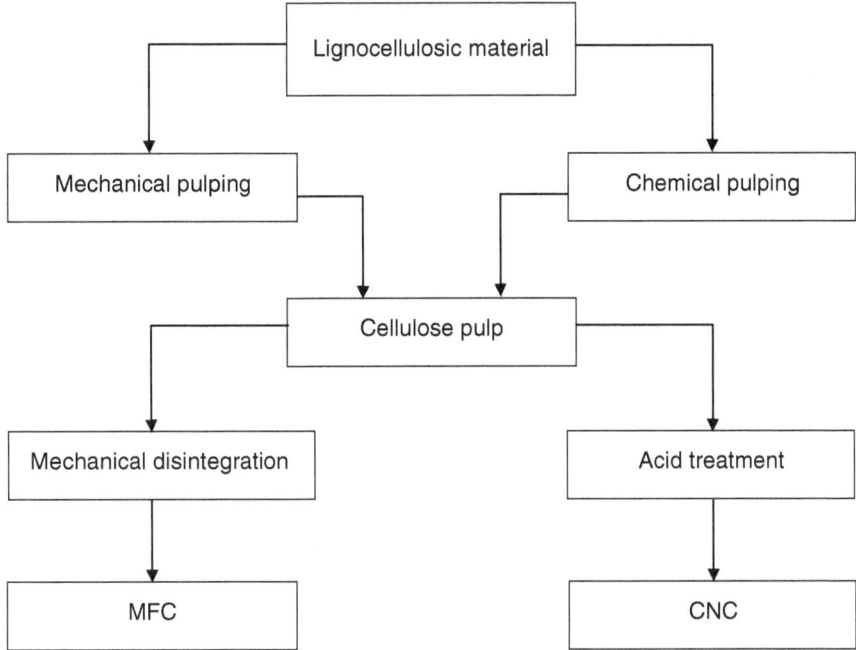

Fig. 1.2 Schematic representation of the different processes to obtain cellulose pulp following microfibrillated and nanocrystalline nanocellulose

nanocrystals (CNC) by acid treatment with sulfuric acid (Saini et al. 2014). It is also possible using enzymatic methods or a combination of them (Kalia et al. 2011).

In fact, the nano arrangement of the cellulose molecules described before, provides the isolation of nanofibrils from initial fibers by breaking up the glycosidic bonds and cleaving the interfibrillar contacts in the disordered domains (amorphous regions) of nanofibrils.

Figure 1.3, shows the methodology used to obtain nanocrystalline cellulose. In this approach, the acid hydrolysis (with H_2SO_4) was used to induce the dissolution of the amorphous regions of the fibres and preserving their highly-crystalline structure after filtration, centrifugation and dialysis.

Figure 1.4, shows the procedure used by Khalil et al. (2014) to individualize cellulose nanofibers using a combination of chemical pre-treatment and high-intensity ultrasonication.

Apart from its vegetable origin, cellulose is also produced by a family of sea animals called tunicates, algae and some aerobic nonpathogenic bacteria (Klemm et al. 2005). Bacteria like *Glucanacetobacter genus* biosynthesized a very pure natural exopolysaccharide known as bacterial cellulose (BC). It is produced in the form of a 3D network swollen gel (~90% water) of nano- and microfibrils (10–100 nm width). BC presents unique properties including high mechanical

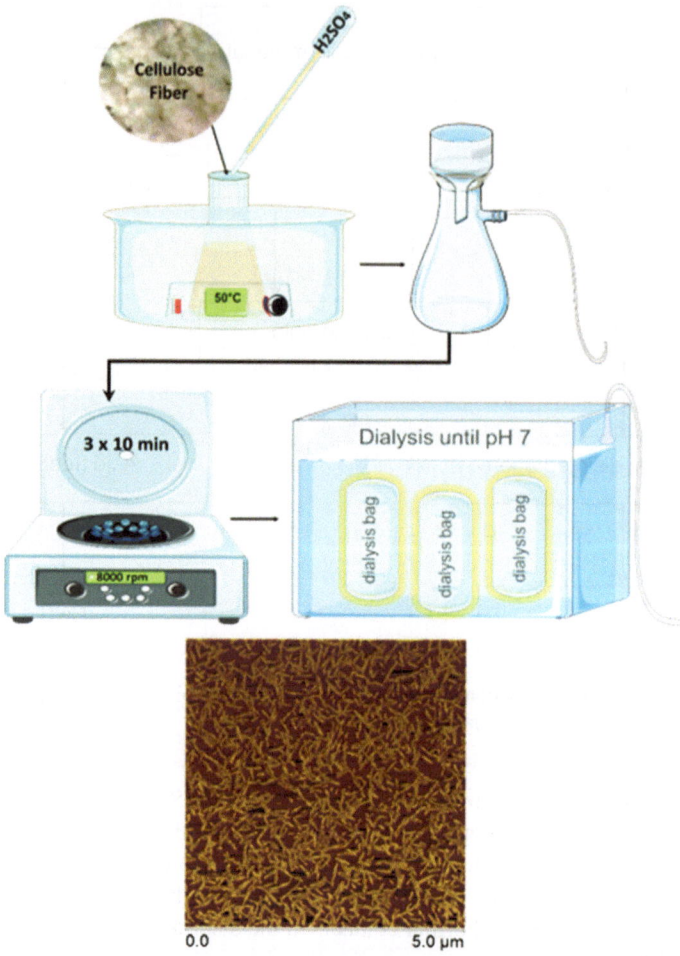

Fig. 1.3 Chemical procedure consisting in converts large pulp fibres (cm) into small nanocellulose (nm): nanocrystalline cellulose. AFM image of cellulose nanocrystals

strength, high crystallinity, high water holding capacity, biocompatibility and high porosity, been a type of nanostructured cellulose widely used as a biomaterial (Rajwade et al. 2015).

1.2.2 Functional Modification of Nanostructured Cellulose

As already mentioned, along cellulose' chain several hydroxyl groups (primary C-6 and secondary C-3) are available and can be used for modification through chemical reactions. The chemical modification of these groups gives rise to new functional

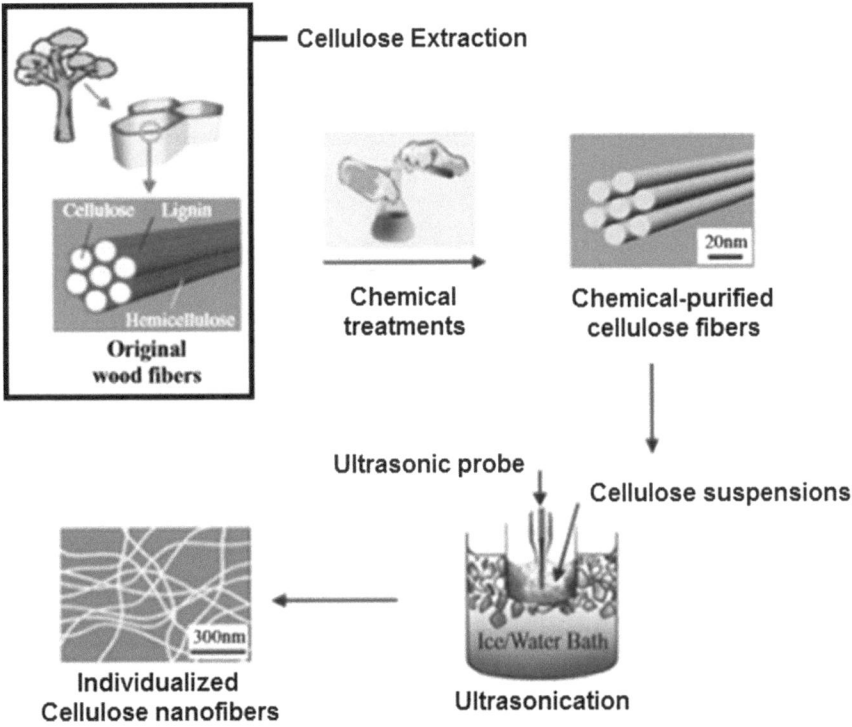

Fig. 1.4 Cellulose nanofibers individualization using a combination of chemical pre-treatment and high-intensity ultrasonication. Adapted with the authorization of Abdul Khalil et al. (2015)

materials that are described in this section and all along this book. Nanostructured cellulose materials can be used as such or modified. As nanocellulose is often used as reinforcing or active agents the modifications are generally carried out on the surface of the nanocrystals or nanofibrils. The functionalization can be used to improve the biological properties of the nanostructured cellulose and also to alter the affinity with water, since cellulose is naturally hydrophobic being this property not always desired. Thus these modifications will prevent self-aggregation and promote efficient dispersion in non-aqueous media (Bajpai 2017). At the same time it is important to preserve the original morphology of the nanocrystals. Figure 1.5 lists some examples of chemical modification on the surface of the cellulose nanocrystals.

Regarding the functionalization of nanostructured cellulose directed for medical applications, different approaches and biomolecules have been used to modify nanocellulose (Fig. 1.6). For instance, functional nanoparticles with fluorescent labelling capacity could be prepared by covalently graft fluorescent molecules (e.g. FITC, fluorescein-5'-isothiocyanate) on the surface of nanostructured cellulose (Dong et al. 2007). The obtained materials can have a potential use in optical bioimaging and biosensors. Another example of nanocellulose modification for

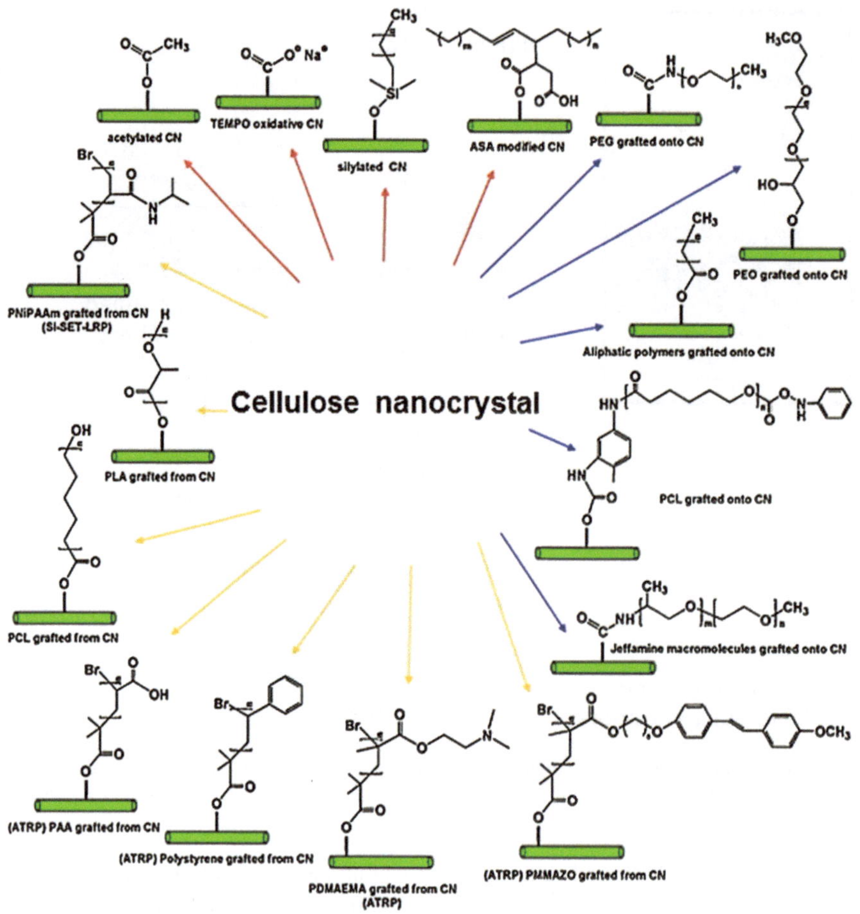

Fig. 1.5 General chemical modifications of cellulose nanocrystals (CNCs). Abbreviations: *PEG* polyethylene glycol, *PEO* polyethylene oxide, *PLA* polylactic acid, *PAA* polyacrylic acid, *PNiPAAm* poly(N-isopropylacrylamide, *PDMAEMA* poly(*N,N*-dimethylaminoethyl). Adapted with the authorization of Lin et al. (2012)

medical application, in particular for nanoplatelet gels, is the grafting of amine-terminated monomers onto surface-modified cellulose nanocrystals followed by click chemistry (Filpponon and Argyropoulos 2010). Initially, the primary hydroxyl groups on the surface of the nanocrystals were selectively activated by converting them to carboxylic acids by the use of TEMPO-mediated hypohalite oxidation. After, carbodiimide-mediated procedure was carried out between precursors with an amine functionality and the carboxylic acid group. The click chemistry reaction, the Cu(I)-catalyzed Huisgen 1,3-dipolar cycloaddition between the azide and the alkyne, surface-activated nanocrystals was employed and nanocrystalline materials in a unique regularly packed arrangement were obtained.

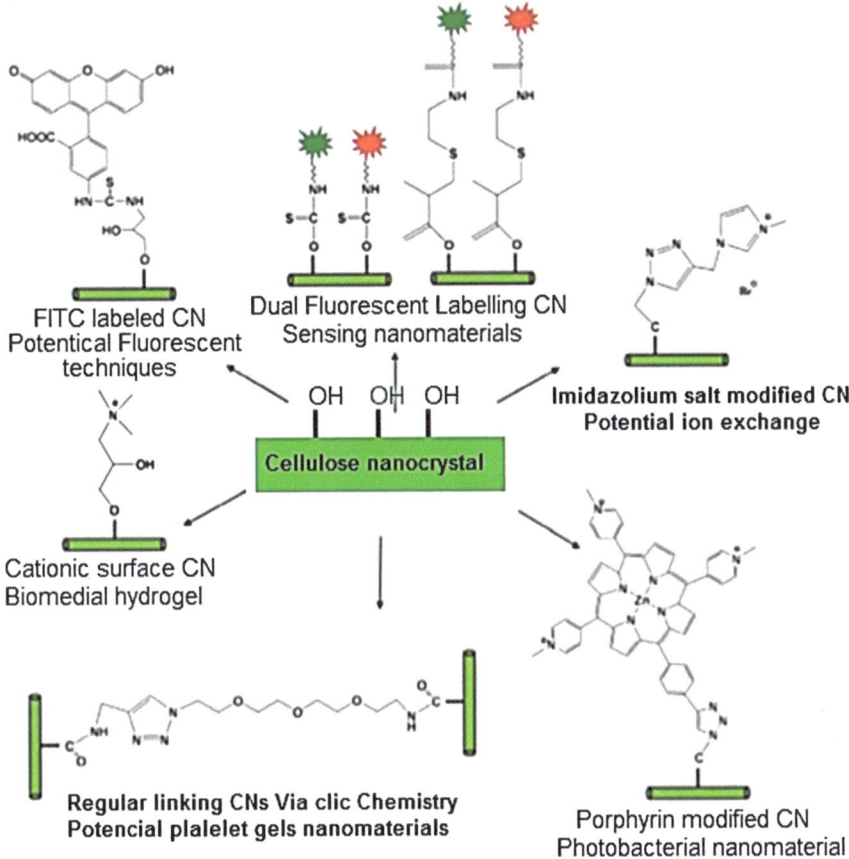

Fig. 1.6 Chemical modifications of cellulose nanocrystals (CNCs) for potential applications in medical and biotechnological sectors. Adapted with the authorization of Lin et al. (2012)

1.2.2.1   Making Antimicrobial Nanostructured Cellulose Biomaterials

Cellulose itself does not present antimicrobial activity, thus nanostructured cellulose-based antimicrobial biomaterials are achieved by the: (i) conjunction (chemical approach) of antimicrobial bioactive agents on the surface of the nanocellulose materials or (ii) incorporation (physical approach) of antimicrobial inorganic (e.g. silver particles (Ag) and its derivatives) and organic (e.g. nanochitin) agents.

For instance, cellulose nanocrystals were modified with allicin and lysozyme through carbodiimide cross-linking procedure to evaluate their antimicrobial properties (Jebali et al. 2013). Their activity was evaluated by the microdilution method and the results showed that unmodified CNCs had few antimicrobial activities, but allicin-conjugated CNCs and lysozyme-conjugated CNCs had good antifungal and antibacterial effects against standard strains of *Candida albicans, Aspergillus niger, Staphylococcus aureus,* and *Escherichia coli.*

Carpenter et al. (2012) have investigated the photobactericidal activity of cellulose nanocrystals modified with a porphyrin-derived photosensitizer (Fig. 1.6, Photobacterial material). The capacity of the synthesized porphyrin-cellulose-nanocrystals to mediate bacterial photodynamic inactivation was investigated as a function of bacterial strain, incubation time and illumination time. The modified nanostructured cellulose material showed excellent efficacy toward the photodynamic inactivation of *Acinetobacter baumannii*, multidrug-resistant *Acinetobacter baumannii* and methicillin-resistant *Staphylococcus aureus.*

Antimicrobial and biocompatible bacterial cellulose-based membranes were prepared by grafting aminoalkyl groups at the surface of BC membranes (Fernandes et al. 2013a). The aminoalkyl-grafted bacterial nanocellulose membranes were prepared in three stages: (1) hydrolysis of the silane derivative (3-aminopropyltrimethoxysilane, APS); (2) adsorption of the hydrolyzed species onto bacterial cellulose fibrils; and (3) chemical condensation reaction. The obtained membranes showed to be destructive against Gram-positive bacterium (*S. aureus*) and Gram-negative bacterium (*E. coli*). Moreover, the membranes also showed to be nontoxic to human adipose-derived mesenchymal stem cells.

In a different manner, silver nanoparticles or nanochitin have been used as bioactive agents to turn nanostructured cellulose in antimicrobial-based materials. Martins et al. (2012) have developed nanocomposites constituted by nanofibrillated cellulose (NFC) and Ag nanoparticles using electrostatic assembly approach via polyelectrolytes as macromolecular linkers between NFC and Ag nanoparticles. The NFC/Ag nanocomposites showed to be resistance towards *S. aureus* and *K. pneumoniae* microorganisms when compared with NFC modified by polyelectrolytes linkers without Ag nanoparticles. More recently, Robles et al. (2016) claimed that the incorporation of chitin nanocrystals in cellulose nanofibers-based materials has a positive effect on the inhibition growth of *Arpergillus sp.* fungus in the final materials. The authors associated this behavior with the presence of residual NH_2 groups on the surface of the chitin nanocrystals.

1.3 Properties of Nanostructured Cellulose as a Biomaterial

Nanocellulose presents unique physical, mechanical and biological properties allowing applications in most varied areas like nanocomposite materials, packaging, paper technology, cosmetics, and medicine.

1.3.1 Physical, Mechanical and Thermal Properties

Compared with conventional cellulose fibres, nanocellulose exhibit high surface area, low density, and excellent mechanical strength (Lin and Dufresne 2013). Thus, nanocellulose presents good pre-requisites to be used as reinforcement in composites to improve the mechanical properties of the host material. The Young's modulus of cellulose material is influenced by crystallinity and the interaction of amorphous fraction and crystalline regions (Cabrera et al. 2011). For instance, the theoretical value of Young's modulus for the native cellulose perfect crystal was estimated to be 167.5 GPa (Tashiro and Kobayashi 1991).

Electrospinning as post-treatment can improve the properties of electrospun nanofibers being considered a simple process that can be used in different applications (Joshi et al. 2015). The authors developed cellulose nanocrystals in poly-caprolactone nanofibers and studied the effects on crystallinity, mechanical strength, biocompatibility, and biomimetic mineralization of the material. The cellulose used in the study was regenerated from cellulose acetate using the post-electrospinning treatment. The study presented good results in mechanical properties and wettability. The research presented different techniques to obtain a material with superior properties, and suggested a material suitable for packaging or coating surface applications.

The thermal and optical performance of nanocomposites based on polysimethyl-siloxane (PDMS) and cellulose nanocrystals (CNCs) as bio-based nanofillers can be improved. To do so, Planes et al. (2016), functionalized the surface of CNCs to promote the dispersion of the hydrophilic CNCs in the hydrophobic polymer matrix (PDMS). Functionalizations were based on acylation of cellulose nanocrystals with vinyl esters, hydrolysis/condensation of methyl trimethoxysilane and hydrosilylation with hydride terminated poly (dimethylsiloxane).

1.3.2 Biological Properties

The biocompatibility and low cytotoxicity, biodegradability and bioreabsorbability properties of major biopolymers, offer many advantages as biomaterials when compared to ceramic and metal materials used for medical applications (Riva et al. 2015).

Thus, nanostructured cellulose has gained increasing interest not only because of its intrinsic physical and mechanical properties but also due to its biological properties, *i.e.* biodegradability, biocompatibility, and low cytotoxicity.

1.3.2.1   Biocompatibility

Biocompatibility is considered the absence of inflammatory, cytotoxicity or invasive response in native cells, tissues, or organs in vivo. Thus, one of the requirements for biomedical applications is that the material must be biocompatible.

The biocompatibility of cellulose-based materials has been investigated and demonstrated. Jia et al. (2013) prepared electrospun composite scaffolds based on microcrystalline cellulose (MCC) and cellulose nanocrystals and claimed that these materials considerably improved the cell viability and morphology on vascular smooth muscle cell. The authors suggested that MCC providing anchors for cells grow within the 3D network of scaffolds and cellulose nanocrystals improve the cell adhesion. This suggests that nanostructured cellulose materials are promising materials for tissue engineering applications to improve the biocompatibility of the final biomaterial.

An in vivo study of subcutaneous bacterial cellulose implantation in rats, showed that after 12 weeks no fibrotic capsule or giant cells were detectable by microscopy. This is a clear indication of no foreign body reaction in the animals (Helenius et al. 2006).

Moreover, recently some studies have been developed to enhance nanocellulose's biocompatibility.

Nitrogen-containing plasma was used to improve the cell affinity of the bacterial cellulose membranes. Pertile et al. (2010) found that the nitrogen plasma-treated BC presented an increasing of adhesion and proliferation of endothelial and neuroblast cells.

Wang et al. (2013) grafted zwitterionic carboxybetaine brushes from cellulose membrane via Activator Regenerated by Electron Transfer ATRP (ARGET-ATRP) to improve blood compatibility. The data demonstrated that cellulose membrane had good blood compatibility performed on lower platelet adhesion and protein adsorption without causing hemolysis.

1.3.2.2   Low Cytotoxicity

In general, there is no indication of serious injury (cytotoxicity, inflammatory effects) of nanostructured cellulose on in vitro cellular or genetic level and on in vivo animal tests. Nonetheless, the number of studies regarding the cytotoxicity of cellulose Nano forms (*i.e.* cellulose nanocrystals, nanofibers and bacterial cellulose) is still limited. Recently, few review articles concerning the current knowledge on the biological impact of nanostructured cellulose have been published (Camarero-Espinosa et al. 2016; Endes et al. 2016; Roman 2015). Roman (2015) performed a literature revision regarding the cytotoxicity of cellulose nanocrystals (CNCs) in pulmonary, oral, dermal tissues. The review draws attention to the contradictory results presented in the literature and the need for further studies related to the potential adverse health effects of CNCs by various exposure routes. The author described the effects of CNCs as a function of physicochemical properties (such as surface chemistry, particle charge, degree of aggregation and particle size), which governs their interactions with the tissue and cell.

The cytotoxicity of CNCs was tested against 9 different cell lines using two different tests—3-(4,5-dimethylthiazol-2-yl)-2,5-diphenyltetrazoliumbromide (MTT) and lactate dehydrogenase (LDH) assays. No cytotoxic effects were observed against the 9 cell lines in the concentration range from 0 to 50 µg/mL of CNCs over 48 h (Dong et al. 2012). In another study, Martin et al. (2011), showed that aerosolized CNCs may induce some respiratory toxicity and inflammatory effects on 3D human lung cells. The risk is regarding the inhalatory exposure under high concentrations of released CNC powders.

In vitro studies with 3T3 fibroblast cells using the test of cell membrane, cell mitochondrial activity and DNA proliferation, demonstrated no toxic phenomena in two types of morphology of pure cellulose nanofibers (CNF): thin and dense structures and porous structures. This work involved direct and indirect contact between the materials and the cells (Alexandrescu et al. 2013).

In microfibrillated cellulose materials, no inflammatory effects or cytotoxicity on mouse macrophage and human monocyte were found, only low acute environmental toxicity (Vartiainen et al. 2011).

Bacterial cellulose (BC) is considered the most biocompatible material in the family of nanostructured cellulose associated to its degree of purity. To date no cytotoxicity of BC was observed in previous studies in vitro on osteoblast, L929 fibroblast cells and human umbilical vein endothelial cells (Chen et al. 2009; Jeong et al. 2010).

1.3.2.3 Biodegradability

Biodegradability and bioabsorbability are the ability of the material and its products to degrade and/or be absorbed or safety eliminated from the body.

The degradation of cellulose occurs via hydrolysis by cellulase enzymes that hydrolyze its β-1,4 D-glucose linkages. Cellulases are made by many types of fungus and bacteria, but are not present in animals. Because of the lack of cellulases in the human body, there are no mechanisms for the large-scale breakdown of cellulose. Thus, cellulose may be considered as non-biodegradable in vivo or really slowly degradable, this could be ideal for some applications, but a problem for others.

Oxidized cellulose has shown to be more susceptible to hydrolysis and therefore potentially degradable by the human body. Consequently, currently, multiple works on improving nanocellulose's biodegradability through oxidation have been done through periodate oxidation in vitro (Li et al. 2009; Luo et al. 2013; Czaja et al. 2014) and in vivo conditions (Czaja et al. 2014).

1.4   Nanostructured Cellulose for Biomedical Applications

As demonstrated, nanostructured cellulose and its derivatives have various advantages and properties as a biomaterial and have been widely used in biomedical, pharmaceutical, and cosmetic industries as carrier systems for delivery, tissue engineering, wound care, healthcare, etc. (Czaja et al. 2006; Thomas 2008; Chang and Zhang 2011; Fernandes 2013b).

Human and plant tissues, such as bone and cartilage and wood, are structured at a nanometric scale and exhibit a hierarchical structure up to the macroscale. Their morphological similarities enable the exploitation of lignocellulosic materials in the development of nanostructured composites targeting delivery systems and tissue engineering and regeneration (Fernandes 2013b). For instance, microcrystalline cellulose (CMC) is widely used in pharmaceutics, mainly as a binder/diluent in oral tablets and capsule formulations in which are used in wet-granulation and direct-compression. In addition, CMC is used in cosmetics and food products. Microcrystalline cellulose is not absorbed systemically following oral administration and thus has little toxic potential (Rowe et al. 2009). Similar examples of cellulose derivatives used in pharmaceutical applications are sodium carboxymethylcellulose, silicified microcrystalline cellulose, cellulose acetate and cellulose acetate phthalate (Rowe et al. 2009).

The versatility of the cellulosic materials in biomedical, cosmetic and biotechnological applications is illustrated in Fig. 1.7.

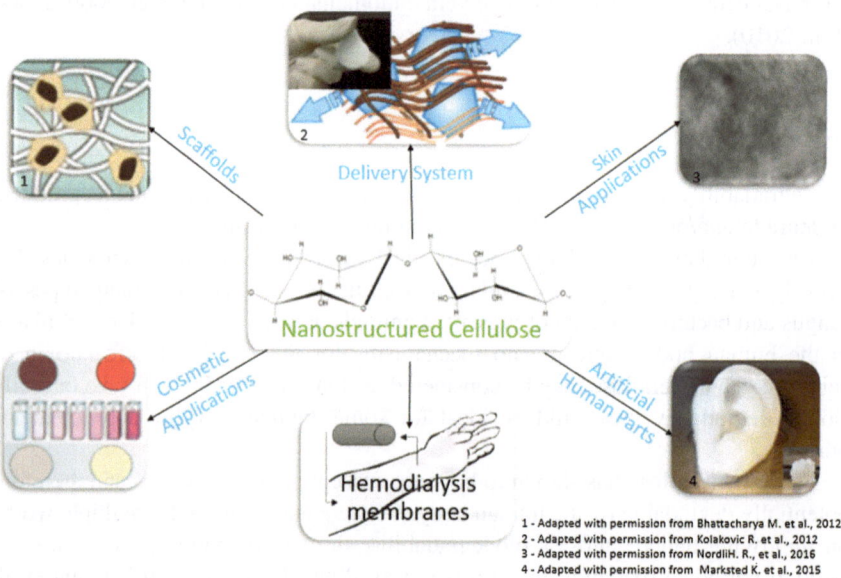

1 - Adapted with permission from Bhattacharya M. et al., 2012
2 - Adapted with permission from Kolakovic R. et al., 2012
3 - Adapted with permission from NordliH. R., et al., 2016
4 - Adapted with permission from Marksted K. et al., 2015

Fig. 1.7 Examples of nanostructured cellulose as a biomaterial in: scaffold materials; delivery systems; skin applications; artificial human body parts; hemodialysis membranes; and cosmetic applications

1.4.1 Nanostructured Cellulose-Based Delivery Systems

There are several studies reporting nanocellulose materials in delivery systems, where the functional substance (e.g. drug, protein, enzymes and hormones) used in the system (films, membranes, tablets, gels) and the strategy to place the substance into the cellulose material were evaluated (Maver et al. 2015; Valo et al. 2013; Kolakovic et al. 2012; Barbosa et al. 2017).

1.4.1.1 Films/Membranes

Nanostructured cellulose in the form of films, membranes and composites has been used for as drug delivery and as a barrier materials.

Maver et al. (2015) showed the potential of thin films of cellulose as a platform for delivery of diclofenac. The film was obtained by spin-coating after mixing trimethylsilyl cellulose (DSSi:2.5) with diclofenac dissolved in tetrahydrofuran (Maver et al. 2015).

Kolakovic et al. (2012) reported the application of nanofibrillated cellulose as a matrix-former material for drug delivery. Indomethacin, itraconazole, and beclomethasone were introduced during the cellulose film production by a filtration method. The authors concluded that nanofibrillar cellulose is a promising material for drug delivery.

A cellulose-based superabsorbent polymer composites (SAPCs) was developed by Anirudhan and Rejeena (2014) as a potential drug delivery vehicle. Sawdust of *Mangifera indica* was used to obtain cellulose nanoparticles. The composite poly (acrylic acid-co-acrylamide-co-2-acrylamido-2-methyl-1-propanesulfonic acid)-grafted nanocellulose/poly (vinyl alcohol) was evaluated for the in vitro gastrointestinal release of amoxicillin. The results showed that this composite could be a great vehicle for the in vitro administration of amoxicillin into the gastrointestinal tract.

The study of Mohanta et al. (2014) reported the use of nanocrystalline cellulose for hydrophobic drug delivery. The methodology used, represented in Fig. 1.8, was a layer-by-layer self-assembly approach for the fabrication of multilayer thin films and microcapsules using nanocrystalline cellulose (NCC) and chitosan (CH). The materials were developed as anticancer drug delivery systems using doxorubicin hydrochloride and curcumin (a water-insoluble drug). The results demonstrated that the microcapsules were successfully loaded with doxorubicin but not with curcumin.

Barbosa et al. (2016), first obtained cellulose nanocrystals (CNC) from flax fibers varying times and temperatures into a sonication bath, and then, the authors developed a membrane with CNC and chlorhexidine (CHX), with different amounts of the drug (0.015, 0.0015 and 0.00015 g). The obtained membrane system showed barrier properties like antimicrobial efficiency against *Staphylococcus Aureus*. Regarding the antimicrobial activity, the results showed that the CNC membrane

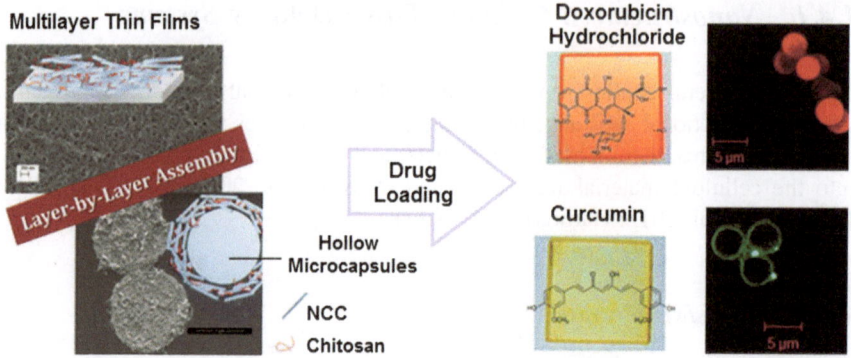

Fig. 1.8 Illustration of drug (doxorubicin hydrochloride and curcumin) loading using multilayer thin films and microcapsules based on chitosan and nanocrystalline cellulose. Adapted with the authorization of Mohanta et al. (2014)

with 0.015 g of CHX inhibited total bacterial growth after 1 and 24 h. The authors concluded that the quantities delivered were maintained over at least 48 h even with a low concentration of drugs. According the study, the CNC with CHX membrane showed good results as a biomaterial capable to control bacteria growth.

1.4.1.2　Tablet Excipient

Kolakovic et al. (2011) evaluated the potential for administration of nanofibrillated cellulose as a tablet excipient material. The authors used spray-dried nanofibrillated cellulose that was compared with two commercial grades of microcrystalline cellulose. Tablets made of nanofibrillated cellulose powder and its mixtures with microcrystalline cellulose with or without paracetamol were produced by direct compression and after wet granulation. The results showed a faster drug release from tablets made by direct compression of nanofibrillated cellulose, due the disintegration behaviour of these nanocellulose forms. The authors concluded that nanofibrillated cellulose could be used as an excipient for tablet production.

1.4.1.3　Aerogels and Hydrogels

Aerogels and hydrogels have been used as a drug carrier due to its ease manufacture and application. Different combinations of nanostructured cellulose and other polymers to provide mechanical stability and biological compatibility have been prepared into hydrogel formulations to evaluate their potential as a drug delivery system.

Valo et al. (2013) studied the behaviour of cellulose nanocrystals aerogels prepared by freeze-dry in gas templates for the development of drug nanoparticles.

Beclomethasone dipropionate (BDP) nanoparticles were integrated into the aerogels. The study concludes that this material is versatile and useful in controlled drug delivery.

Lin and Dufresne (2013) obtained supramolecular hydrogels formed with modified cellulose nanocrystals (Fig. 1.9). The grafting efficiency of β-cyclodextrin and the introduction of pluronic on the surface of cellulose nanocrystals showed a significant improvement of structural and thermal stability of hydrogels. The in situ supramolecular hydrogels were used as drug release for in vitro release of doxorubicin and showed the behaviour of prolonged drug release.

Lin et al. (2016) produced double membrane hydrogels based on sodium alginate (anionic) with cationic cellulose nanocrystals to increase structural stability and control the delivery of drugs (Fig. 1.10a). Through variations in the composition the properties of the airways were altered, varying the effects of drug release. Thus, the material exhibited a special double-membrane structure and alternative drug delivery behaviours that demonstrated potential use in oral and curatives administration in biomedical applications. The mechanism of drug release for the double-membrane hydrogel is depicted in Fig. 1.10b.

Wang et al. (2016) reports the use of nanocrystalline cellulose (NCC) as substrate to produce silver nanoparticles (AgNPs) at room temperature for antimicrobial barrier. Glucose was applied like reducing agent. To prepare the NCC the authors used sulfuric acid hydrolysis of cotton pulp. The Fig. 1.11 shows the methodology of the NCC-assisted generation of AgNPs. The generated AgNPs assisted by NCC was used for developing a visible, sensitive, and quantitative glucose assay. This colorimetric and non-enzymatic assay for glucose detection shows a wide linear range from 0.116 μM to 0.4 mM. The AgNPs/NCC also was evaluated against Gram-negative (*Escherichia coli* and multi-drug resistance

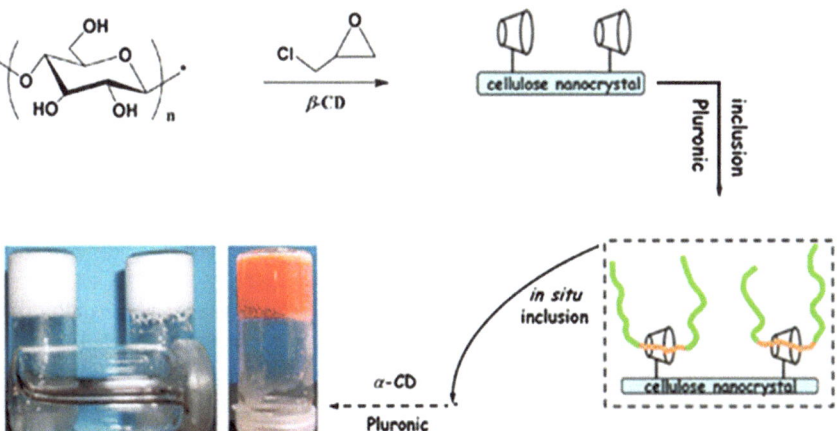

Fig. 1.9 Scheme of hydrogels development by the grafting of β-cyclodextrin and introduction of pluronic polymers on the surface of cellulose nanocrystals. Adapted with the authorization of Lin and Dufresne (2013)

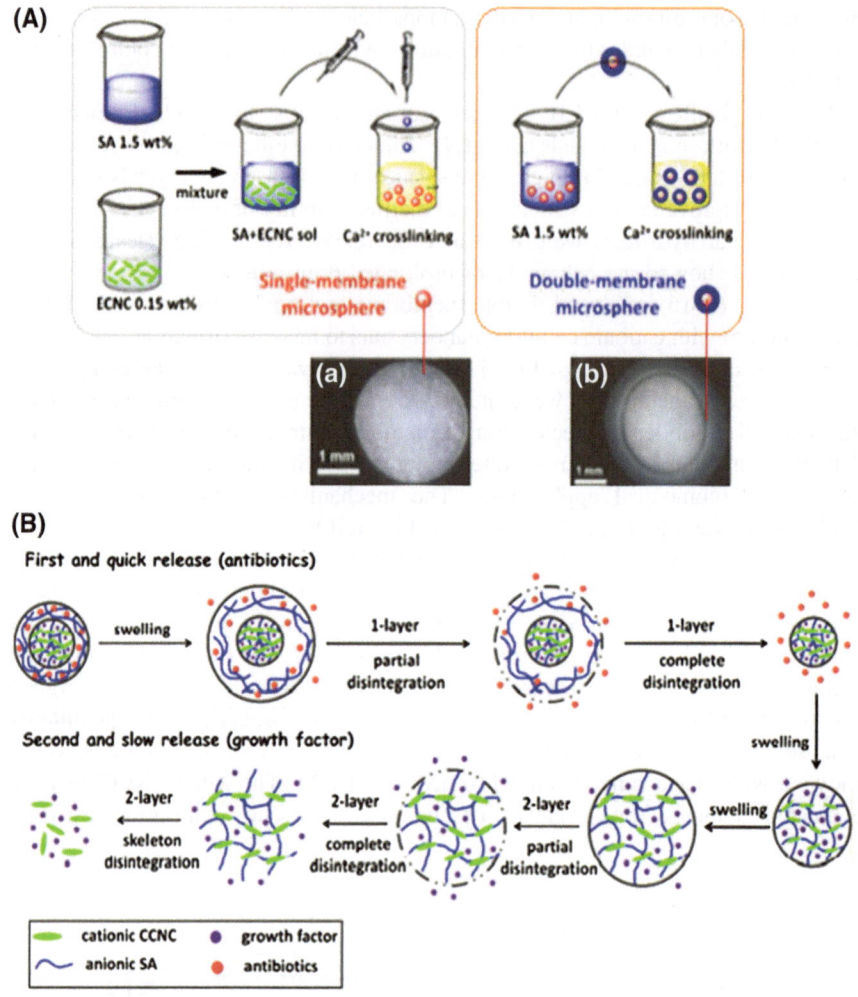

Fig. 1.10 a Methodology to obtain double membrane hydrogels from cationic cellulose nanocrystals and anionic alginate. **b** Drug release model for the double-membrane hydrogel (formation of cationic chemically modified cellulose nanocrystals and anionic alginate under the pH 7.4 condition). Adapted with authorization of Lin et al. (2016)

Escherichia coli) and Gram-positive (*Staphyloccocus aureus* and methicillin-resistant *Staphylococcus aureus*). The results exhibit an enhanced antibacterial activity of the AgNPs/NCC for both Gram-negative and Gram-positive bacteria in comparison with the commercial AgNPs. These results supports the potential use of AgNPs/NCC in clinical diagnosis, environmental monitoring, and the control of bacteria.

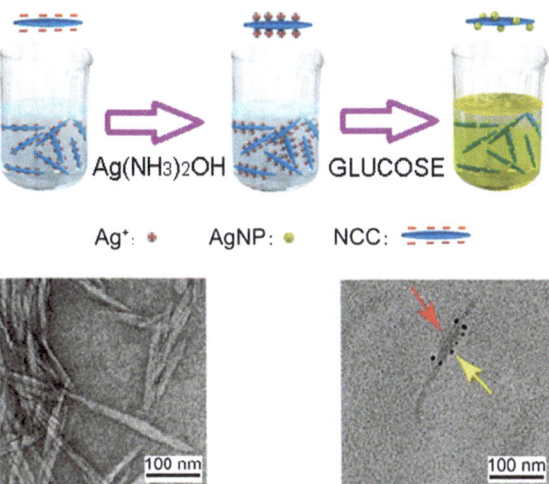

Fig. 1.11 Schematic illustration of generated AgNPs assisted by NCC. First of all, the NCC was dispersed in aqueous solution. After the Ag(NH₃)₂OH solution was add and silver ions absorbed onto the surface of NCC and AgNPs were obtain from redox reaction between glucose and silver ions (*up*). TEM images (*down*) of dispersed NCC (*left*) and of obtained AgNPs assisted by NCC (*right*): NCC (*red arrow*) and generated AgNPs (*yellow arrow*). Adapted with authorization Wang et al. (2016)

1.4.2 Nanostructured Cellulose-Based Systems for Tissue Engineering

1.4.2.1 Scaffolds for Cell Culture Growth

Tissue engineering (TE) investigates technologies for regeneration of functional living tissues and organs.

Nanofibrillar cellulose (NFC) hydrogels in the form of 3D-cell culture scaffolds were developed by Bhattacharya et al. (2012). In their study, the authors used commercial cell culturing hydrogels as reference, and showed that NFC scaffolds stimulated hepatocyte (HepaRG and HepG2 human hepatic cell lines) cell spheroid formation without the addition of bioactive components (Fig. 1.12). The authors emphasized that the raw material used was formed by a single component, NFC hydrogel, which shows the versatility in the use of nanostructured cellulose (Bhattacharya et al. 2012).

Naseri et al. (2016), proposed the tailoring of mechanical properties using nanocellulose. It was expected to find alternatives to tissue engineering via 3D porous structures, which could act as templates for the formation of new tissues and act as guidance for cell growth, facilitating nutrient and oxygen transport. Furthermore, the scaffolds' high porosity, high phosphate buffered saline uptake

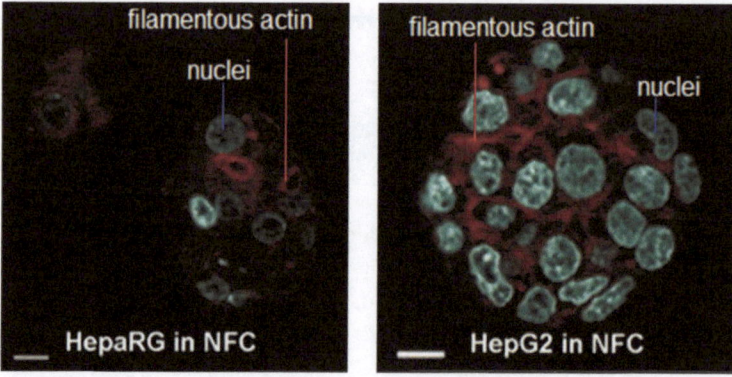

Fig. 1.12 NFC supported cell spheroid formation using HepaRG and HepG2 human hepatic cell lines: confocal microscopy images with structural staining of filamentous actin (*red*) and nuclei (*blue*). Adapted with the authorization of Bhattacharya et al. (2012)

and good cytocompatibility towards chondrocytes, could drive significant benefits to cell attachment and extracellular matrix formation.

Joshi et al. (2015) reported a natural-synthetic hybrid of cellulose acetate (CA) and polycaprolactone (PCL) fabricated by electrospinning process. The CA in the hybrid fibre was transformed into cellulose (CL) by post-electrospinning treatment via alkaline saponification. This process is demonstrated in Fig. 1.13. The material was characterized by confocal microscopy and scanning electron microscopy and the cell viability was evaluated. The results showed excellent ability of cell proliferation and growth for MC3T3 and the authors suggested the use of the composite scaffolds in tissue regeneration applications.

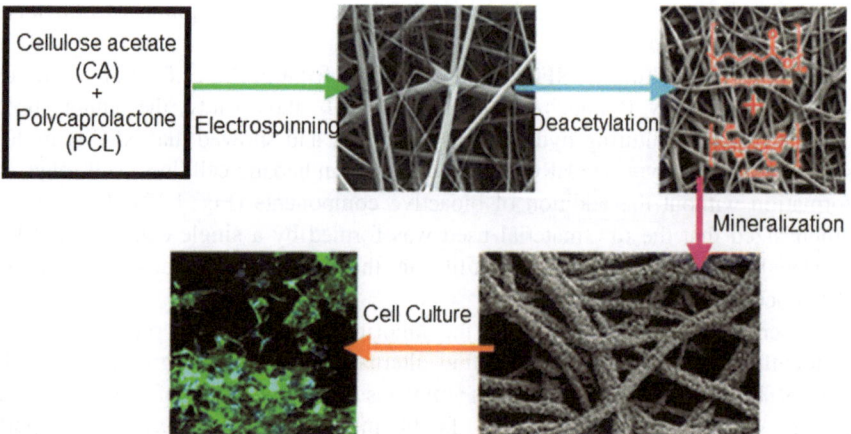

Fig. 1.13 Images of the material after the different techniques used in the work. Adapted with the authorization of Joshi et al. (2015)

1.4.2.2   Nanocellulose in Skin Purposes

Skin is the largest organ of the human body, and nanostructured cellulose, including bacterial cellulose (Pittella and Porto 2015), has been considered a promising material for skin applications namely as wound dressing or for skin replacement (Leonida and Kumar 2016).

A desirable characteristic of materials used in skin recovery is the ability to absorb the exudate during the healing process, and its removal from a wound surface after recovery (Fu et al. 2013). The hydroquinone (HDQ) is well-known by the properties of inhibition of melanin and discoloration of skin. Some studies suggest the incorporation of HDQ in a suitable carrier such as cellulose nanocrystal could improve the therapeutic effect on skin wound (Taheri and Mohammadi 2015).

Brown et al. (2015), hold a United States Patent (US 8,951,551 B2) regarding nanostructured cellulose obtained from different sources and its application in wound dressing. Advantages of the nanostructured cellulose cited by authors include a non-allergenic nature, simple processing, in addition to promotes natural host cellular migration to a wound site. The authors quoted its application at donor sites and in medium thickness wounds. The material was created for the use in a novel wound healing system, and more particularly as a wound dressing for a wide variety of wound types, locations, shapes, depth and stage(s) of healing.

Nordli et al. (2016) developed a nanofibrillated cellulose aerogels for skin repair from *Pinus radiata* pulp fibres using sodium hydroxide followed by TEMPO-mediated oxidation (Fig. 1.14). The study showed that the material is a

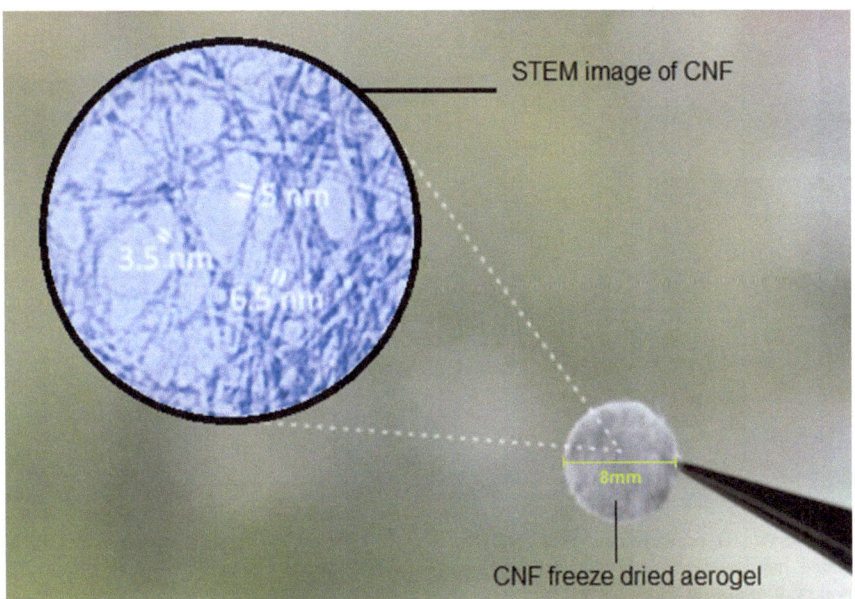

Fig. 1.14 CNF freeze dried aerogel (8 mm punch biopsy sample) and SEM image of its morphology. Adapted with the authorization of Lin and Nordli (2016)

good candidate to be used like wound dressing. The biomaterial improved the absorption of water and showed good results in cytotoxicity.

Singla (2017) reported the application of cellulose nanocrystals extracted from bamboos with low concentrations of silver in the production of wound dressings for the skin repair. The nanocrystals were extracted from the leaves of two species of bamboo *Dendrocalamus hamiltonii* and *Bambusa bambos* through chemical and mechanical processes. Studies in vivo of the obtained material showed good water absorption capacity—that keeps the wounded tissue moist, strong antibacterial activity and rapid epithelial recovery, avoiding inflammatory processes and increasing the proliferation of fibroblasts.

Bacterial cellulose membranes are used in the treatment of severe burns and the composites reinforced with cellulose nanostructured show beneficial effect for skin tissue repair (Leonida and Kumar 2016). In general, skin applications of nanocellulose frequently reported in the literature are obtained from bacterial cellulose.

1.4.2.3   Nanocellulose for Cartilage, Bone and Tendons

Nanofibrillated cellulose (NFC) produced by mechanical refinement and enzymatic treatment was used with alginate to obtain a bioink for 3D bioprinting of living soft tissues with cells (process based on fast cross-linking ability of alginate). The material was developed with the necessary characteristics for to be printable in a 3D printing and useful for anatomically shaped cartilage structures, such as a human ear and sheep meniscus, were 3D printed using MRI and CT images as blueprints (see Fig. 1.15) (Markstedt et al. 2015).

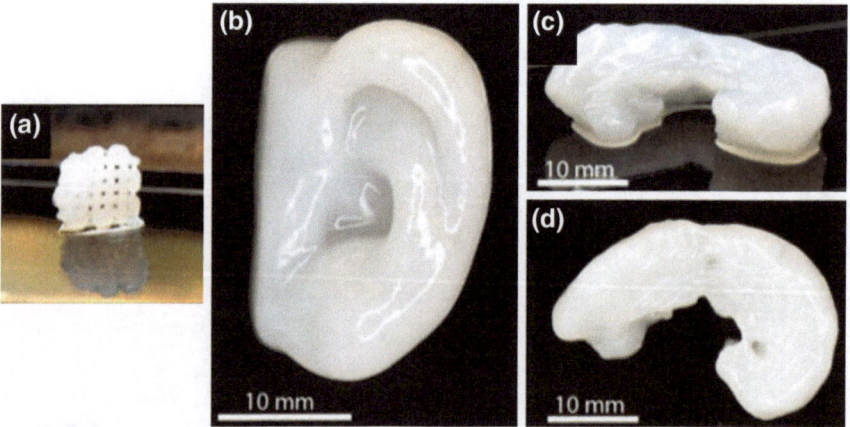

Fig. 1.15 a 3D printed small grids (7.2 × 7.2 mm) with Ink 8020 after cross-linking; **b** 3D printed human ear; and **c, d** sheep meniscus with Ink 8020. Adapted with authorization of Markstedt et al. (2015)

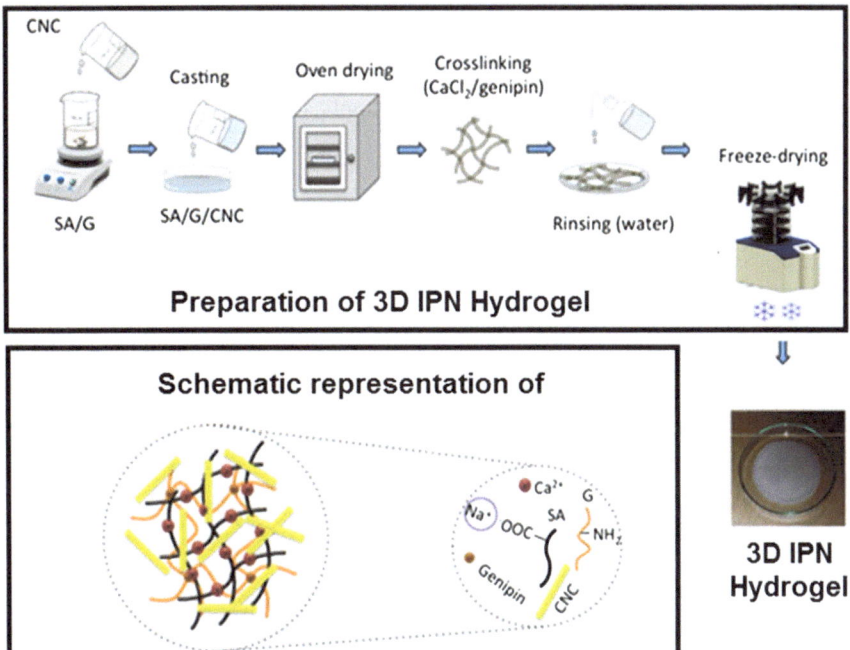

Fig. 1.16 Preparation of hydrogels and 3D IPN hydrogel picture. Adapted with authorization of Naseri et al. (2016)

Naseri et al. (2016) prepared three-dimensional hydrogels with pores interconnected from reticulated polymer network of sodium alginate reinforced with cellulose nanocrystals. The material was prepared via freeze-drying and stabilized using $CaCl_2$ and genipin, as show in Fig. 1.16. In addition to high porosity, the material presented nanostructured roughness, which facilitates cell growth and adhesion. Also, the material presented adequate mechanical properties compared with natural cartilage.

Zhou et al. (2013) prepared bio-nanocomposite scaffolds reinforced with cellulose nanocrystals (CNCs) using maleic anhydride (MAH) grafted poly (lacticacid) (PLA) as matrix. Cotton-based CNCs were obtained using 64% sulfuric acid aqueous hydrolysis followed by high-pressure homogenization. The addition of CNCs improved the thermal stability and mechanical properties of MPLA/CNC composites. The bio-composite scaffolds were non-toxic to human adult adipose derived mesenchymal stem cells (HASCs) and capable of supporting cell proliferation. The authors conclude that the material could be potentially suitable in the bone tissue engineering and highlight the use of CNCs as advanced function materials.

Fragal et al. (2016) explored the use of cellulose nanowhiskers to generate the biomimetic growth hydroxyapatite for bone tissue engineering. Hybrids materials with cellulose nanowhiskers and hydroxyapatite were produced by different inorganic acid hydrolyses to generate cellulose particles with surface groups to induce hydroxyapatite mineralization (Fig. 1.17). The authors conclude that cellulose nanowhiskers can be used as a base for growing hydroxyapatite and this material could be used for bone regeneration applications.

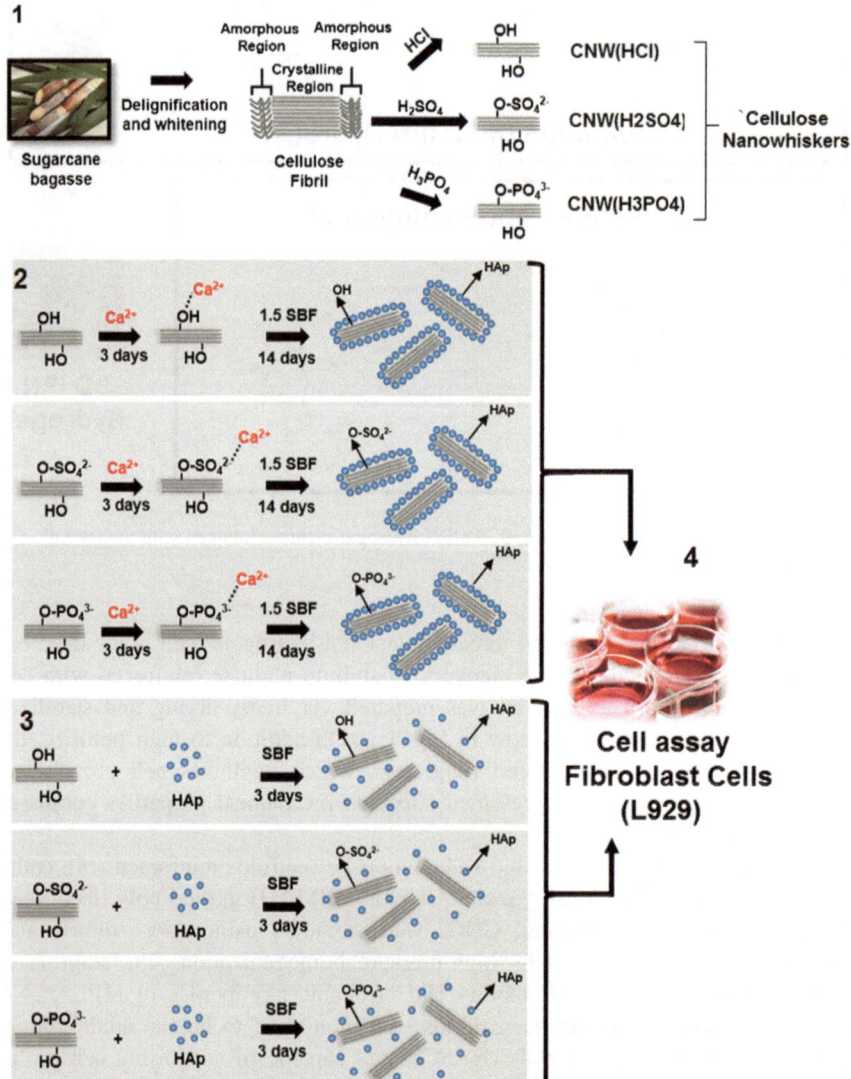

Fig. 1.17 Schematic representation of biomaterial syntheses derived from cellulose nanowhiskers from sugarcane bagasse with HAp using different protocols: biomimetic method with SBF and wet chemical precipitation. (*1*) Cellulose nanowhiskers synthesis. (*2*) Mineralization of HAp on the surface of cellulose nanowhiskers by the biomimetic method. (*3*) Preparation of cellulose nanowhiskers and HAp hybrid material by wet chemical precipitation. (*4*) Cell assay with Fi hybrid M cells (L929). Adapted with the authorization of Fragal et al. (2016)

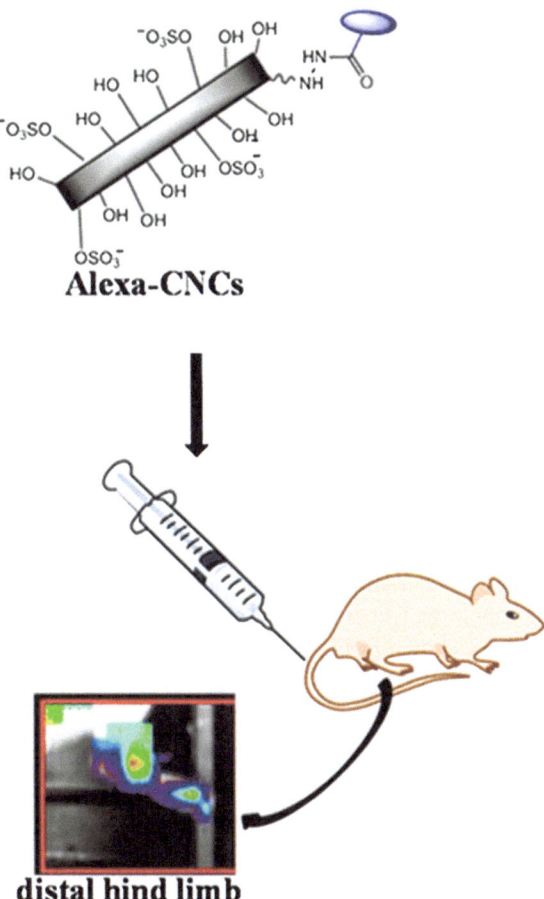

Fig. 1.18 Illustration of surface modification applied in CNCs and performed analyses. Adapted with the authorization of Colombo et al. (2015)

Alexa-CNCs

distal hind limb

Colombo et al. (2015) investigated the interaction of cellulose nanocrystals with living organisms (Fig. 1.18). They performed in vivo and ex vivo analysis and found a chemical interaction between the bone matrix and the CNCs. Making the CNCs a possible nano-device applicable in bone diseases with emphasis on bone tumours.

In 2012, Mathew and co-workers evaluated fibrous cellulose nanocomposite scaffolds prepared by partial dissolution as potential ligament or tendon substitute. The study indicated that the method used represented a viable route to develop biomaterials with good mechanical properties and cell compatibility required for medical applications (Mathew et al. 2012). In 2013, the same authors claimed to have developed collagen-based cellulose nanofiber reinforced materials for the same applications: artificial ligament/tendons. The nanocomposite demonstrated in simulated body conditions the strength and elongation, which is the extension of natural ligament or tendon. Cytocompatibility test indicated that the nanocomposites allowed adhesion, growth, and differentiation of human ligament cells and human endothelial cells.

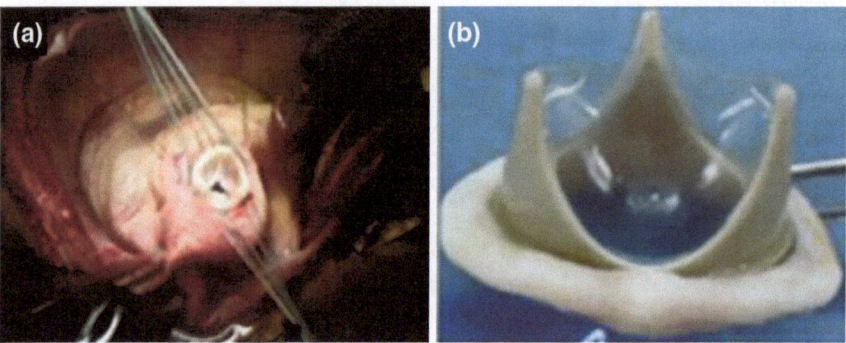

Fig. 1.19 Nanocellulose–polyurethane prosthetic heart valve: **a** valve implant, **b** heart valve.
Adapted with authorization of Cherian et al. (2011)

1.4.2.4 Artificial Blood Vessels, Heart Valve Prosthesis, Vessel Patches

As previously mentioned, cellulose can be used in different ways. An example is the
study of Cherian and collaborators (2011) where nanocellulose extracted from
pineapple leaf fibers was embedded in polyurethane for the fabrication of heart
valvules shown in Fig. 1.19 (Cherian et al. 2011). The authors concluded that the
mechanical performance of composite with 5 wt% cellulose was optimal, and the
nanocomposites were capable to be used in diverse biomedical applications,
including scaffolds, cardiovascular implants, repair of auricular cartilage, vascular
grafts, urethral catheters, mammary prostheses, penile prostheses, adhesion barriers
and artificial skin.

He et al. (2014) obtain a cellulose nanocomposite nanofibers reinforced with
cellulose nanocrystals via electrospinning. The method (see Fig. 1.20) provided a
uniform morphology of the electrospun cellulose/CNCs nanocomposite nanofibers.
The characterizations carried out suggested that the nanocomposite can be used as a
blood vessel.

1.4.2.5 Artificial Cornea

Wang et al. (2010) investigated the use of bacterial cellulose (BC) as an artificial
cornea replacement. BC was formed by nano-sized fibril network and the hydrogel
composite of BC/poly(vinyl alcohol) (BC/PVA) were synthesized. Results such as
high light transmittance, increased mechanical properties and good thermal properties indicated that the BC/PVA composites showed promising characteristics as
artificial cornea composite material.

Fig. 1.20 Scheme of the experimental procedure of cellulose nanocomposite nanofibers reinforced with cellulose nanocrystals. Adapted with authorization of He et al. (2014)

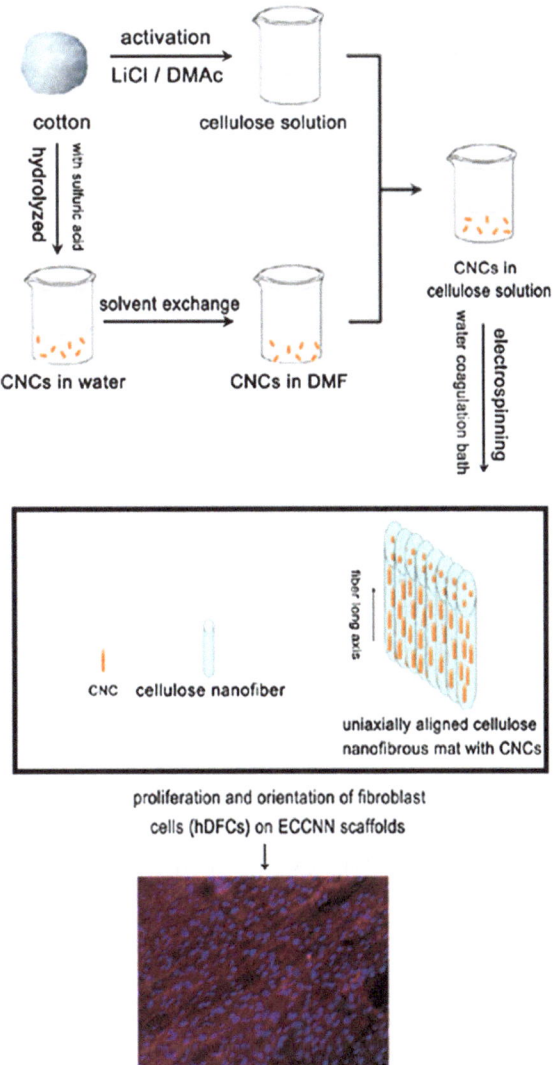

1.4.2.6   Artificial Urethra

Nanofibrilated cellulose from pineapple leaf fibers was used as fillers into polyurethane to obtain a polymer composites by Cherian et al. (2011). The study proved that the reinforcement was efficient and the composites were utilized to develop some medical implants. The authors showed the possibility of the use of these composite as urethral catheters, highlighting the bioresorbability and biodegradability of the material.

1.5   Nanostructured Cellulose-Based Systems for Other Applications

1.5.1   Hemodialysis Membranes

A study presenting the development of a composite used to purify blood was presented at the Euromembrane conference in 2012. Carlsson et al. obtained a nanocomposite with polypyrrole coated on nanofibrilated cellulose and investigated the material as hemodialysis membranes. Results showed that the pore sizes of the composite can be adapted to fit into blood purification applications.

1.5.2   Cosmetic Applications

Nanostructured cellulose can be used in health care applications such as cosmetics (Ioelovich 2008; Miller 2015). For instance, Hu et al. (2016) demonstrated dry cellulose nanocrystals' potential as stabilizers of emulsions and their potential for redisperse dried emulsions in water (Fig. 1.21).

1.5.3   Technical Applications

The excellent properties of nanocellulose encouraged its exploitation in many areas, including materials such as sensors or biosensors. The response of the sensor is quantified through changes of its properties, when subjected to certain stimuli. Many biosensors have been developed thanks to the incorporation of peptides or enzymes by covalent linkages, salt bridges or physical interactions in the nanocellulose materals. An example is a human neutrophil elastase (HNE) sensor. HNE is a protease related to immune cells under inflammatory processes and has been shown to be a specific

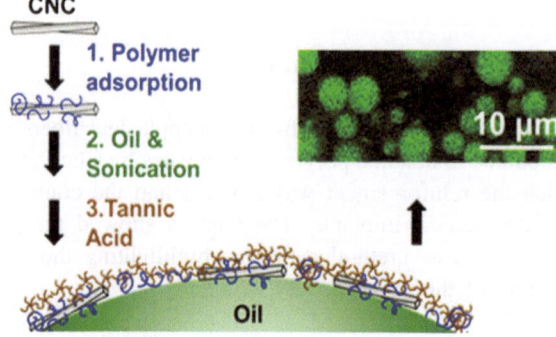

Fig. 1.21 Scheme to make dried and re-dispersible cellulose nanocrystals for emulsions. Adapted with the authorization of Hu et al. (2016)

marker for many diseases (Edwards et al. 2013). The authors reported biosensor using in a colorimetric detection based on HNE tripeptide substrate, n-Succinyl-Alanine–Alanine-Valine-paranitroanilide (Suc-Ala–Ala-Val-pNA) covalently attached to glycine esterified cotton cellulose nanocrystals.

Nanocellulose can also be used in nanocomposite to immobilize enzymes. Incame et al. (2013) demonstrated that nanocrystalline cellulose with gold nanoparticles placed by cationic polyethylenimine (PEI) can immobilize glucose oxidases (GOx).

Other work reported a highly stretchable resistive nanopaper applied in strain sensors. The nanopaper was prepared by filtering a mixture of nanocellulose fibril with graphene (weight ratio 1:1). In this report, the nanocellulose was considered an efficient, low-cost, and green "binder" to increase the processability of crumpled graphene. The composite showed to be an efficient human-motion detection sensor (Yan et al. 2014).

Navarro et al. (2016), obtained nanocellulose by extraction of wood pulp and applied it as fluorescent markers by controlled radical polymerization, as shown in Fig. 1.22. Thus, the material proved to be a viable marker enabling the optics detection when applied to live juvenile Daphnis.

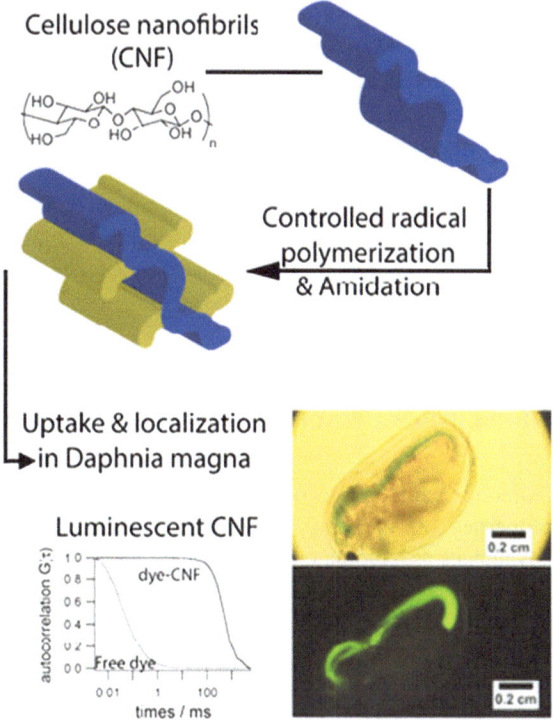

Fig. 1.22 Schematic illustration of the conversion of cellulose nanofibrils (CNF) into fluorescently CNF. Adapted with the authorization of Navaro et al. (2016)

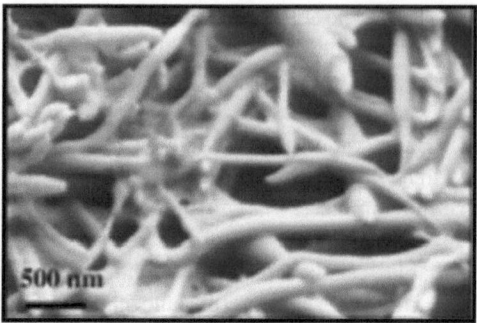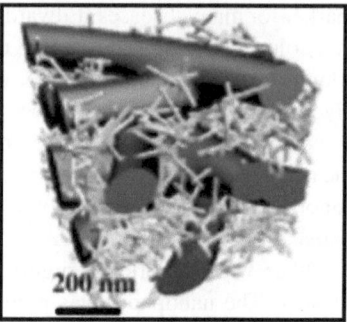

Fig. 1.23 SEM image of cellulose nanowhisker-modified PAN electrospun nanofibrous scaffold (*above*). Schematic representations of electrospun nanofibrous scaffolds: infused nanowhiskers forming loose cross-linked mesh (*left*); and nanowhiskers collapsed onto the scaffold, forming bundles (*right*). Adapted with the authorization of Ma et al. (2011)

A membrane system was developed by impregnating cellulose nanowhiskers into an electrospun polyacrylonitrile (PAN) nanofibrous scaffold supported by a poly ethylene terephthalate (PET) substrate (Fig. 1.23). This multilayered nanofibrous microfiltration (MF) composite exhibited higher adsorption capacity to a positively charged dye over a nitrocellulose-based MF commercial membrane (Ma et al. 2012). In addition, the novel membrane also showed full retention capability against two different bacteria (*E. coli* and *B. diminuta*), thus log reduction value (LRV) of 2 against virus model (bacteriophage MS2) as compared with an LRV of 1 for the analogues commercial material GS0.22, suggesting a suitable membrane to liquid flux treatment system.

Thus, nanocellulose has great potential for technological applications, due to the simplicity modification, either by inserting proteins, enzymes, functional groups or metallic nanoparticles. Several technical applications will emerge such as biodegradable sensors or food packaging with quality indicators.

1.6   Future Perspectives and Final Remarks

Different techniques used to obtain nanocellulose confer different characteristics and properties on the final material providing to nanostructured cellulose various advantages as a biomaterial. The examples described in this book represent selected

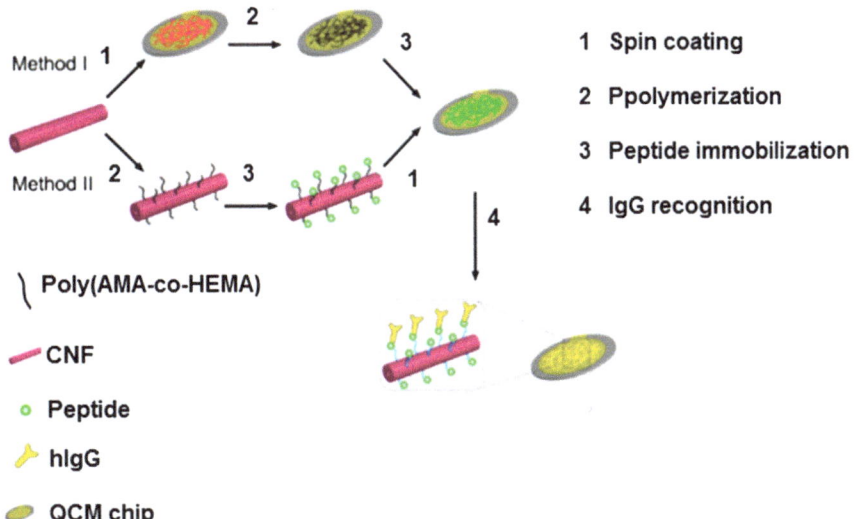

1   Spin coating

2   Ppolymerization

3   Peptide immobilization

4   IgG recognition

\ Poly(AMA-co-HEMA)

— CNF

o Peptide

hIgG

QCM chip

Fig. 1.24 Schematic illustration of the preparation of peptide-CNF films for hIgG detection using two Different methods. Adapted with the authorization of Zhang et al. (2013)

applications of nanostructured cellulose as delivery systems and as tissue engineering systems in the biomedical field. The successful demonstration of nanocellulose as a biomaterial has lead to the development of new nanostructured cellulose-based materials by new technologies like 3D bioprinting (see Sect. 1.4.2.3.) developed by Markstedt et al. (2015) for 3D bioprinting living soft tissue with cells. A different approach represented in Fig. 1.24, describe the preparation of bioactive films by the conjugation of a short peptide onto modified nanofibrilated cellulose. The study has demonstrate a new platform and a disposable CNF-based sensor or nanopaper for detection of human immunoglobulin G (Zhang et al. 2013).

Dugan et al. (2010) prepared by spin-coating cellulose nanowhiskers (CNWs). The spin-coating was used to produce surfaces of CNWs with a high degree of radial orientation on pieces of optical glass pre-treated with the cationic polyelectrolyte polyallylamine hydrochloride (PAHCl). They produced submonolayers of sparsely adsorbed tunicin CNWs (Fig. 1.25). By the methodology used, this work got a high degree of orientation without building up multiple layers of polyelectrolyte. Mechanical, physical and biological characterization of CNWs surface was performed. The data suggested that sub monolayer surfaces of tunicin CNWs could produce a high degree of orientation using a spin-coating method. Myoblasts have

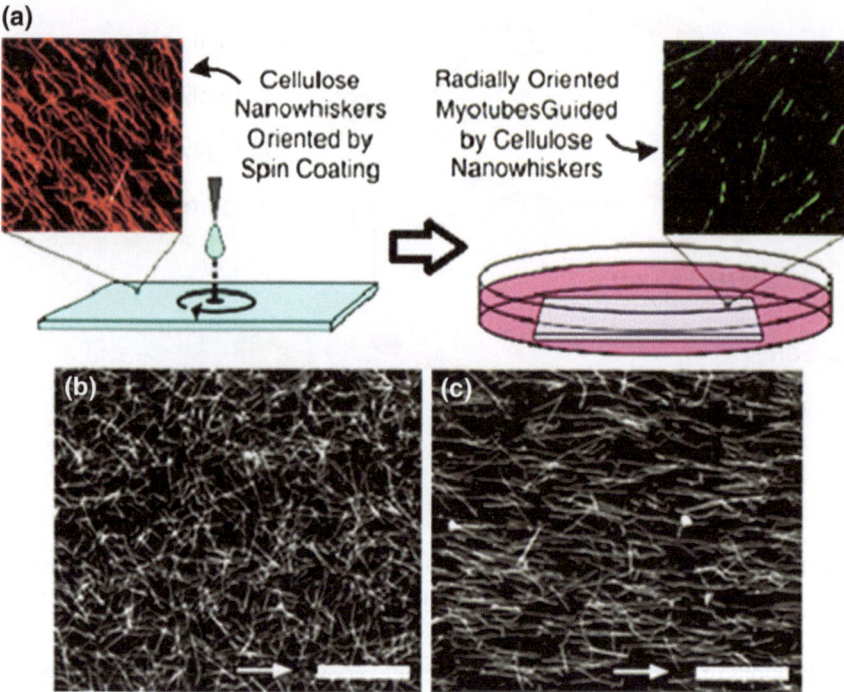

Fig. 1.25 a Radially oriented submonolayer surfaces of CNWs prepared by spin-coating; **b, c** AFM topography image of sample surface obtained by spin-coating with different conditions. Adapted with the authorization of Dugan et al. (2010)

been shown to sense the topography of these surfaces effectively and to orientate relative to the bulk direction of CNW orientation, being suggested a potential use of CNWs for tissue engineering applications.

Lewis et al. (2016) reported the hydrothermal gelation of suspensions of aqueous cellulose nanocrystal (CNC). A variation in the temperature, time, and CNC concentration was used in the preparation. The CNC hydrogels were prepared using an aqueous suspension of CNC-H+ into a Teflon-lined stainless steel autoclave. The emphasis in this methodology is the hydrothermal treatment, when the desulfation of the CNC surface occurs, resulting in a destabilization of CNCs in suspension causing gelation (Fig. 1.26). The obtained hydrogels were investigated by rheology, electron microscopy, and spectroscopy. The authors suggest that this CNC hydrogels can be applied in drug delivery, insulation, and as tissue scaffolds. Moreover, the material can be easily used as aerogel after dry for different applications.

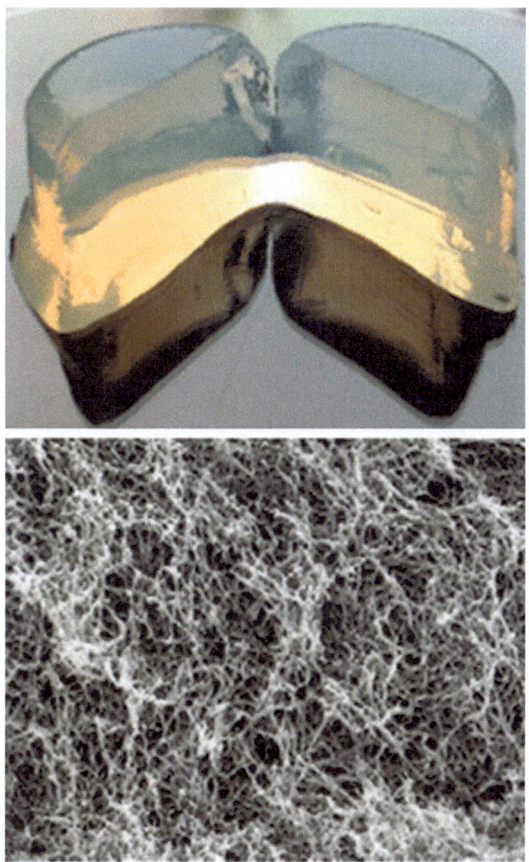

Fig. 1.26 Optical photograph of 2 wt% CNC-H+ suspension (120 °C for 20 h) (*Up*). CNCs desulfation forming a gel (*Down*). Adapted with the authorization of Lewis et al. (2016)

The possibility of modifying the cellulose nanoparticles expands its applications and makes their study a very valuable subject. In this way, our research group developed nanocomposites films with differents types and size particles of cellulose materials, an exemplo is the cellulose acetate with hidroxiapatite film (Fig. 1.27). This is a very promising work because this film can be used in a wide range of applications. Another very positive point is the biomass used for obtaining the cellulose acetate, the banana stem that was treated as described by Noremberg et al. (2017), through simple physical and chemical processes.

With all these selected examples, nanostructured cellulose has proved to be a very versatile material that can be used in different ways to make biomaterials.

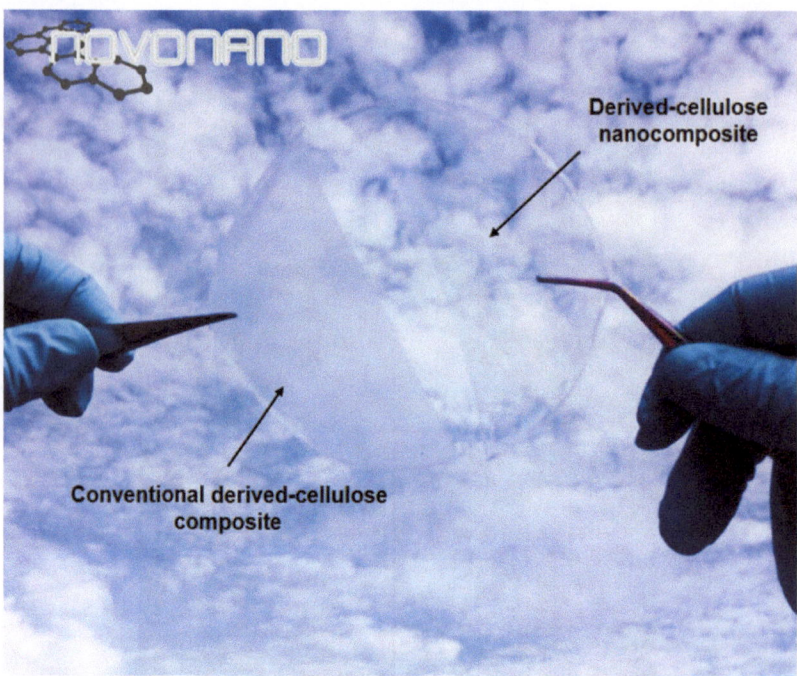

Fig. 1.27 Nanocomposite film from banana stem cellulose acetate: and hidroxiapatite nanoparticles (*right*), hidroxiapatite microparticles (*left*)

Thus, the evolution achieved with nanocellulose in recent years brings numerous optimistic perspectives in relation to future applications. Nanocellulose-based biomaterials are expected to become a useful matrix substance for various biomedical applications in the future.

References

Abdul Khalil HPS, Bhat AH, Bakar AA, Tahir PMd., Zaidul ISM, JawaidM (2015) Cellulosic nanocomposites from natural fibers for medical applications: a review. In: Pandey JK, Takagi H, Nakagaito AN, Kim H-J (eds) Handbook of polymer nanocomposites. Processing, performance and application: Volume C: Polymer nanocomposites of cellulose nanoparticles. Springer, Heidelberg, pp 475–511

Abitbol T, Rivkin A, Cao Y et al (2016) Nanocellulose, a tiny fiber with huge applications. Current Opinion in Biotechnology 39:76–88

Alexandrescu L, Syverud K, Gatti A, Chinga-Carrasco G (2013) Cytotoxicity tests of cellulose nanofibril-based structures. Cellulose 20:1765–1775

Alila S, Besbes I, Rei Vilar M et al (2013) Non-woody plants as raw materials for production of microfibrillated cellulose (MFC): a comparative study. Ind Crops Prod 41:250–259

Anirudhan TS, Rejeena SR (2014) Poly(acrylic acid-co-acrylamide-co-2-acrylamido-2-methyl-1-propanesulfonic acid)-Grafted Nanocellulose/Poly(vinyl alcohol) Composite for the In Vitro Gastrointestinal Release of Amoxicillin. J Appl Polym Sci 131:1–12

Bajpai P (2017) Modification of nanocellulose to improve properties. Bajpai P(ed) Pulp Pap Ind. Elsevier, Amsterdam, pp 91–104

Barbosa AM, Robles E, Ribeiro JS et al (2016) Cellulose nanocrystal membranes as excipients for drug delivery systems. Materials 9:1002

Bhattacharya M, Malinen MM, Lauren P et al (2012) Cellulose hydrogel promotes three-dimensional liver cell culture. J Controll Release 164:291–298

Brown RM Jr, Claja W, Jeschke M et al (2015) Multiribbon nanocellulose as a matrix for wound healing. US Patent 8,951,551 B2 31 Aug 2006

Cabrera RQ, Meersman F, McMillan PF et al (2011) Nanomechanical and structural properties of native cellulose under compressive stress. Biomacromolecules 12:2178–2183

Carlsson DO, Nyström G, Ferraz N et al (2012) Development of nanocellulose/polypyrrole composites towards blood purification. Proc Eng 44:733–736

Carpenter BL, Feese E, Sadeghifar H, Argyropoulos DS, Ghiladi RA (2012) Porphyrin-cellulose nanocrystals: a photobactericidal material that exhibits broad spectrum antimicrobial activity. Photochem Photobiol 88:527–536

Camarero Espinosa S, Endes C, Mueller S, Petri Fink A, Rothen Rutishauser B, Weder C, Clif MJD, Foster EJ (2016) Elucidating the potential biological impact of cellulose nanocrystals. Fibers 4:21

Chang CY, Zhang LN (2011) Cellulose-based hydrogels: present status and application prospects. Carbohyd Polym 841:40–53

Chen YM, Xi T, Zheng Y, Guo T, Hou J, Wan Y et al (2009) In vitro cytotoxicity of bacterial cellulose scaffold for tissue engineered bone. J Bioact Compat Polym 24:137–145

Cherian BM, Leão AL, Souza SF et al (2011) Cellulose nanocomposites with nanofibres isolated from pineapple leaf fibers for medical applications. Carbohydr Polym 86:1790–1798

© The Author(s) 2017
N.L.V. Carreño et al., *Advances in Nanostructured Cellulose-based Biomaterials*,
SpringerBriefs in Applied Sciences and Technology,
DOI 10.1007/978-3-319-58158-3

Colombo L, Zoia L, Violatto MB et al (2015) Organ distribution and bone tropism of cellulose nanocrystals in living mice. Biomacromolecules 16:2862–2871

Czaja W, Krystynowicz A, Bielecki S et al (2006) Microbial cellulose–the natural power to heal wounds. Biomaterials 27:145–151

Czaja W, Kyryliouk D, DePaula CA, Buechter DD (2014) Oxidation of cirradiated microbial cellulose results in bioresorbable, highly conformable biomaterial. J Appl Polym Sci 131:39995

Dong S, Roman M (2007) Fluorescently labeled cellulose nanocrystals for bioimaging applications. J Am Chem Soc 129:13810

Dong S, Hirani AA, Colacino KR, Lee YW, Roman M (2012) Cytotoxicity and cellular uptake of cellulose nanocrystals. Nano Life 02:1241006–1241017

Dugan JM, Gough JE, Eichhorn SJ (2010) Directing the morphology and differentiation of skeletal muscle cells using oriented cellulose nanowhiskers. Biomacromolecules 11:2498–2504

Edwards JV, Prevost N, French A et al (2013a) Nanocellulose-based biosensors: design, preparation, and activity of peptide-linked cotton cellulose nanocrystals having fluorimetric and colorimetric elastase detection sensitivity. Engineering 5:20–28

Edwards JV, Prevost N, Sethumadhavan K et al (2013b) Peptide conjugated cellulose nanocrystals with sensitive human neutrophil elastase sensor activity. Cellulose 20:1223–1235

Endes C, Camarero Espinosa S, Mueller S, Foster EJ, Petri Fink A, Rothen Rutishauser B, Weder C, Clif MJD (2016) A critical review of the current knowledge regarding the biological impact of nanocellulose. J Nanobiotechnol 14:78

Fernandes SCM, Sadocco P, Aonso-Varona A, Palomares T, Eceiza A, Silvestre AJD, Mondragon I, Freire CSR (2013a) Bioinspired antimicrobial and biocompatible bacterial cellulose membranes obtained by surface functionalization with aminoalkyl groups. ACS Appl Mater Interfaces 5:3290–3297

Fernandes EM, Pires RA, Mano JF et al (2013b) Bionanocomposites from lignocellulosic resources: properties, applications and future trends for their use in the biomedical field. Prog Polym Sci 38:1415–1441

Filpponon I, Argyropoulos DS (2010) Regular linking of cellulose nanocrystals via click chemistry: synthesis and formation of cellulose nanoplatelet gels. Biomacromolecules 11:1060–1066

Fragal EH, Cellet TSP, Fragal VH et al (2016) Hybrid materials for bone tissue engineering from biomimetic growth of hydroxyapatite on cellulose nanowhiskers. Carbohydr Polym 152:734–746

Fu L, Zhang J, Yang G (2013) Present status and applications of bacterial cellulose-based materials for skin tissue repair. Carbohydr Polym 92:1432–1442

Helenius G, Bäckdahl H, Bodin A, Nannmark U, Gatenholm P, Risberg B (2006) In vivo biocompatibility of bacterial cellulose. J Biomed Mater Res A 76A:431–438

He X, Xiao Q, Lu C et al (2014) Uniaxially aligned electrospun all-cellulose nanocomposite Nanofibers reinforced with cellulose nanocrystals: scaffold for tissue engineering. Biomacromolecules 15:618–627

Hu Z, Marway HS, Kasem H et al (2016) Dried and redispersible cellulose nanocrystal pickering emulsions. ACS Macro Lett 5:185–189

Incani V, Danumah C, Boluk Y (2013) Nanocomposites of nanocrystalline cellulose for enzyme immobilization. Cellulose 20:191–200

Ioelovich M (2008) Cellulose as a nanostructured polymer: a short review. Bioresources 3: 1403–1418

Jebali A, Hekmatimoghaddam S, Behzadi A, Rezapor I, Mohammadi BH, Jasemizad T et al (2013) Antimicrobial activity of nanocellulose conjugated with allicin and lysozyme. Cellulose 20:2897–2907

Jeong SI, Lee SE, Yang H, Jin YH, Park CS, Park YS et al (2010) Toxicologic evaluation of bacterial synthesized cellulose in endothelial cells and animals. Mol Cellular Toxicol 6: 373–380

Jia B, Li Y, Yang B, Xiao D, Zhang S, Rajulu AV, Kondo T, Zhang L, Zhou J (2013) Effect of microcrystal cellulose and cellulose whisker on biocompatibility of cellulose-based electrospun scaffolds. Cellulose 20:1911–1923

Jorfi M, Foster EJ (2014) Recent advances in nanocellulose for biomedical applications. J Appl Polym Sci 132:1–19

Joshi MK, Tiwari AP, Pant HR et al (2015) In situ generation of cellulose nanocrystals in polycaprolactone nanofibers: Effects on crystallinity, mechanical strength, biocompatibility, and biomimetic mineralization. ACS Appl Mater Interfaces 7:19672–19683

Kalia S, Kaith BS, Kaur I (2011) Cellulose fibers: bio- and nano-polymer composites—green chemistry and technology, 1st edn. Springer, Berlin

Klemm D, Heinze T, Wagenknecht W (1998) Comprehensive cellulose chemistry. Wiley-VCG, Weinheim

Klemm D, Heublein B, Fink HP et al (2005) Cellulose: fascinating biopolymer and sustainable raw material. Angew Chem Int Ed 44:3358–3393

Klemm D, Kramer F, Moritz S et al (2011) Nanocelluloses: a new family of nature-based materials. Angew Chem Int Ed 50:5438–5466

Kolakovic R, Peltonen L, Laaksonen T et al (2011) Spray-dried cellulose nanofibers as novel tablet excipient. AAPS Pharm Sci Tech 12:1366–1372

Kolakovic R, Peltonen L, Laukkanen A et al (2012) Nanofibrillar cellulose films for controlled drug delivery. Eur J Pharm Biopharm 82:308–315

Leonida MD, Kumar I (2016) Bionanomaterials for skin regeneration. Springer International Publishing, Switzerland

Lewis L, Derakhshandeh M, Hatzikiriakos SG et al (2016) Hydrothermal gelation of aqueous cellulose nanocrystal suspensions. Biomacromolecules 17:2747–2754

Li J, Wan YZ, Li LF, Liangm H, Wang JH (2009) Preparation and characterization of 2,3-dialdehyde bacterial cellulose for potential biodegradable tissue engineering scaffolds. Mater Sci Eng C 29:635–1642

Lin N, Dufresne A (2013) Supramolecular hydrogels from in situ host-guest inclusion between chemically modified cellulose nanocrystals and cyclodextrin. Biomacromolecules 14:871–880

Lin N, Huang J, Dufresne A (2012) Preparation, properties and applications of polysaccharide nanocrystals in advanced functional nanomaterials: a review. Nanoscale 4:3274–3294

Lin N, Gèze A, Wouessidjewe D et al (2016) Biocompatible double-membrane hydrogels from cationic cellulosen anocrystals and anionic alginate as complexing drugs co-delivery. ACS Appl Mater Interfaces 8:6880–6889

Luo H, Xiong G, Hu D, Ren K, Yao F, Zhu Y et al (2013) Characterization of TEMPO-oxidized bacterial cellulose scaffolds for tissue engineering applications. Mater Chem Phys 143:373–379

Ma H, Burger C, Hsiao BS et al (2012) Nanofibrous microfiltration membrane based on cellulose nanowhiskers. Biomacromolecules 13:180–186

Markstedt K, Mantas A, Tournier I et al (2015) 3D bioprinting human chondrocytes with nanocellulose-alginate bioink for cartilage tissue engineering applications. Biomacromolecules 16:1489–1496

Martin JD, Clift E, Foster J, Vanhecke D, Studer D, Wick P et al (2011) Investigating the interaction of cellulose nanofibers derived from cotton with a sophisticated 3D human lung cell co-culture. Biomacromolecules 12:3666–3673

Martins NCT, Freire CSR, Pinto RJB, Fernandes SCM, Neto CP, Silvestre AJD et al (2012) Electrostatic assembly of Ag nanoparticles onto nanofibrillated cellulose for antibacterial paper products. Cellulose 19:1425–1436

Mathew AP, Oksman K, Pierron D et al (2012) Fibrous cellulose nanocomposite scaffolds prepared by partial dissolution for potential use as ligament or tendon substitutes. Carbohydr Polym 87:2291–2298

Mathew AP, Oksman K, Pierron D et al (2013) Biocompatible fibrous networks of cellulose nanofibres and collagen crosslinked using genipin: potential as artificial ligament/tendons. Macromol Biosci 13:289–298

Maver T, Maver U, Mostegel F et al (2015) Cellulose based thin films as a platform for drug release studies to mimick wound dressing materials. Cellulose 22:749–761

Miller J (2015) Nanocellulosestate of the industry. Available in: http://www.tappinano.org/media/1114/cellulose-nanomaterials-production-state-of-the-industry-dec-2015.pdf. Accessed 10 Feb 2017

Mohanta V, Madras G, Patil S (2014) Layer-by-Layer assembled thin films and microcapsules of nanocrystalline cellulose for hydrophobic drug delivery. ACS Appl Mater Interfaces 6:20093–20101

Naseri N, Deepa B, Mathew AP et al (2016a) Nanocellulose based interpenetrating polymer network (ipn) hydrogels for cartilage applications. Biomacromolecules 17:3714–3723

Naseri N, Poirier JM, Girandon L et al (2016b) 3-Dimensional porous nanocomposite scaffolds based on cellulose nanofibers for cartilage tissue engineering: tailoring of porosity and mechanical performance. RSC Adv 6:5999–6007

Navarro JRG, Wennmalm S, Godfrey J et al (2016) Luminescent nanocellulose platform: from controlled graft block copolymerization to biomarker sensing. Biomacromolecules 17:1101–1109

Nordli HR, Chinga-Carrasco G, Rokstad AM et al (2016) Producing ultrapure wood cellulose nanofibrils and evaluating the cytotoxicity using human skin cells. Carbohydr Polym 150:65–73

Noremberg BS, Silva RM, Paniz OG et al (2017) From banana stem to conductive paper: a capacitive electrode and gas sensor. Sens Actuators B 240:459–467

Pertile RAN, Andrade FK, Alves C, Gama M (2010) Surface modification of bacterial cellulose by nitrogen-containing plasma for improved interaction with cells. Carbohydr Polym 82:692–698

Pittella CQP, Porto LM (2015) Application of bacterial nanocellulose membranes for epithelial tissue repair. REV.Enf-UFJF 1:223–232

Planes M, Brand J, Lewandowski S et al (2016) Improvement of the thermal and optical performances of protective polydimethylsiloxane space coatings with cellulose nanocrystal additives. ACS Appl Mater Interfaces 8:28030–28039

Rajwade JM, Paknikar KM, Kumbhar JV (2015) Applications of bacterial cellulose and its composites in biomedicine. Appl Microbiol Biotechnol 99:2491

Riva GH, García-Estrada J, Vega B et al (2015) Cellulose–Chitosan nanocomposites—evaluation of physical, mechanical and biological properties, cellulose. In: Fundamental Aspects and Current Trends, Chapter 9

Robles E, Salaberria AM, Herrera R, Fernandes SCM, Labidi J (2016) Self-bonded composite films based on cellulose nanofibers and chitinnanocrystals as antifungal materials. Cellulose 144:41–49

Roman M (2015) Toxicity of cellulose nanocrystals: a review. Ind Biotechnol 11:25–33

Rowe RC, Sheskey PJ, Quinn ME (2009) Handbook of pharmaceutical excipients, 6th edn. Pharmaceutical Press, American Pharmacists Association, Washington, DC

Saini JK, Saini R, Tewari L (2014) Lignocellulosic agriculture wastes as biomass feedstocks for second-generation bioethanol production: concepts and recent developments. Biotech 5:337–353

Singla R, Soni S, Kulurkar PM (2017) In situ functionalized nanobiocomposites dressings of bamboo cellulose nanocrystals and silver nanoparticles for accelerated wound healing. Carbohydr Polym 155:152–162

Siró I, Plackett D (2010) Microfibrillated cellulose and new nanocomposite materials: a review. Cellulose 17:459–494

Sunasee R, Hemraz UD, Ckless K (2016) Cellulose nanocrystals: a versatile nanoplatform for emerging biomedical applications. Expert Opin Drug Deliv 13:1243–1256

Taheri A, Mohammadi M (2015) The use of cellulose nanocrystals for potential application in topical delivery of hydroquinone. Chem Biol Drug Des 86:102–106

Tashiro K, Kobayashi M (1991) Theoretical evaluation of three-dimensional elastic constants of native and regenerated celluloses: role of hydrogen bonds. Polymer 32:1516–1526

Thomas S (2008) A review of the physical, biological and clinical properties of a bacterial cellulose wound dressing. J Wound Care 17:349–352

Valo H, Arola S, Laaksonen P et al (2013) Drug release from nanoparticles embedded in four different nanofibrillar cellulose aerogels. Eur J Pharm Sci 50:69–77

Vartiainen J, Pöhler T, Sirola K, Pylkkänen L, Alenius H, Hokkinen J et al (2011) Health and environmental safety aspects of friction grinding and spray drying of microfibrillated cellulose. Cellulose 18:775–786

Wang J, Gao C, Zhang Y et al (2010) Preparation and in vitro characterization of BC/PVA hydrogel composite for its potential use as artificial cornea biomaterial. Mater Sci Eng C 30:214–218

Wang M, Yuan J, Huang X, Cai X, Li L, Shen J (2013) Grafting of carboxybetaine brush onto cellulose membranes via surface-initiated ARGET-ATRP for improving blood compatibility. Colloides Surf B: Biointerface 103:52–58

Wang S, Sun J, Jia Y et al (2016) Nanocrystalline cellulose-assisted generation of silver nanoparticles for non-enzymatic glucose detection and antibacterial agent. Biomacromolecules 17:2472–2478

Yan C, Wang J, Kang W et al (2014) Highly stretchable piezo resistive graphene–nanocellulose nanopaper for strain sensors. Adv Mater 26:2022–2027

Zhang Y, Carbonell RG, Rojas OJ (2013) Bioactive cellulose nanofibrils for specific human igg binding. Biomacromolecules 14:4161–4168

Zhou C, Shi Q, Guo W, Terrell L et al (2013) Electrospun bio-nanocomposite scaffolds for bone tissue engineering by cellulose nanocrystals reinforcing maleic anhydride grafted PLA. ACS Appl Mater Interfaces 5:3847–3854